THE REBIRTH
OF NATURE

THE REBIRTH
OF NATURE

THE GREENING
OF
SCIENCE AND GOD

RUPERT SHELDRAKE

Park Street Press
Rochester, Vermont

Park Street Press
One Park Street
Rochester, Vermont 05767

Library of Congress Cataloging-in-Publication Data
Sheldrake, Rupert.
    The rebirth of nature : the greening of science and God / Rupert Sheldrake.
        p.   cm.
    Originally published: New York : Batam Books, 1991.
    Includes bibliographical references and index.
    ISBN 0-89281-510-8
    1. Nature—Religious aspects. 2. Nature. I. Title.
    [BL65.N35S44   1994]
    113—dc20   94-2811
                                                                        CIP

Printed and bound in the United States on recycled paper

10 9 8 7 6 5 4 3 2 1

Park Street Press is a division of Inner Traditions International

Distributed to the book trade in the United States by American International Distribution Corporation (AIDC)

Distributed to the book trade in Canada by Publishers Group West (PGW), Montreal West, Quebec

To my wife, Jill

CONTENTS

ACKNOWLEDGMENTS

This book is a result of a long personal quest. I cannot begin to list all the plants, animals, places, people, traditions, and ideas that have helped me along the way. I can only express my general sense of gratitude for all that I have been given in the countries in which I have lived—England, the United States, Malaysia, and India—and in the course of my travels in Europe, North America, Asia, and Africa.

Many conversations with friends and colleagues have contributed to this book. Some have taken place in the context of informal visits, some at conferences and symposia, and some in several series of meetings that I attended over the last decade: in particular, the regular gatherings in Cambridge of the Epiphany Philosophers, a group to which I have belonged since 1966; in meetings of the British Scientific and Medical Network; in annual Councils at the Ojai Foundation, California from 1984 to 1987; and at series of small invitational conferences at the

Esalen Institute, California; at Hollyhock Farm, on Cortes Island, British Columbia; at the Institute of Noetic Sciences, Sausalito, California; and at the International Center for Integrative Studies in New York.

In particular, I would like to thank the following people for discussions that have helped shape the content of this book: Ralph Abraham, David Abram, William Anderson, Eric Ashby, Lindsay Badenoch, Robert Bly, David Bohm, Fritjof Capra, Bernard Carr, Christopher Clarke, Paul Davies, Larry Dossey, Lindy Dufferin and Ava, Dorothy Emmet, Warwick Fox, Adele Getty, Edward Goldsmith, Brian Goodwin, David Griffin, Bede Griffiths, Joan Halifax, David Hart, Rainer Hertel, Mae-wan Ho, Francis Huxley, Rick Ingrasci, Colleen Kelley, David Lorimer, Terence McKenna, Ralph Metzner, John Michell, Namkhai Norbu, Robert Ott, the late Michael Ovenden, Nigel Pennick, Anthony Ramsay, Martin Rees, Jeremy Rifkin, Janis Roze, Kit Scott, Ronald Sheldrake, Paolo Silva e Souza, John Steele, Dennis Stillings, John Sullivan, Harley Swiftdeer, Brian Swimme, Robin Sylvan, Peggy Taylor, George Trevelyan, Piers Vitebsky, Lyall Watson, Rex Weyler—and above all, my wife, Jill Purce, to whom this book is dedicated.

I am especially grateful to those who have read various drafts of this book for their helpful comments and criticisms: Lindsay Badenoch, Christopher Clarke, Adele Getty, Bede Griffiths, Francis Huxley, Kit Scott; my British editors, Erica Smith and Kelly Davis, and my American editor, Leslie Meredith, of Bantam Books.

The writing of this book was partially assisted by a grant from the Institute of Noetic Sciences, of which I am a Fellow.

I thank Keith Roberts and his assistants for doing the drawings in Figures 5.1, 5.2, 5.3, 5.4, 6.2, and 7.1, and thank the following for permission to reproduce illustrations: the Trustees of the British Museum (Fig. 1.1); the British Library (Figs. 1.2 and 2.2); Clive Hicks (Fig. 2.1); Ralph Abraham (Fig. 4.2); Oxford University Press (Fig. 5.4); and J. Bloxham and D. Gubbins (Fig. 7.2).

INTRODUCTION

My grandmother came from a family of willow growers in Nottinghamshire, producing osiers for local wickerworkers. My most vivid image of the rebirth of nature came to me when I was staying at the old family farmhouse in Farndon, a village on the river Trent near my hometown, Newark. I was about four or five years old. Near the house, I saw a row of willow trees with rusty wire hanging from them. I wanted to know why it was there, and asked my uncle, who was nearby. He explained that this had once been a fence made with willow stakes, but the stakes had come to life and turned into trees. I was filled with awe.

I forgot all about this incident until a few years ago when it came to mind in a moment of sudden illumination. First of all, there was the memory itself, the moment of insight as I saw how the stakes had turned into living trees. Then came the amazing realization that it summed up much of my scientific career. For over twenty years, in Cambridge, Malaysia, and India, I did research on the development of plants. I was continually fascinated by the interplay between death and regeneration. In particular, I discovered that the plant hormone auxin, which stimulates growth and development and induces the rooting of cuttings, is produced by dying cells.[1] For example, it is produced by the wood cells, which "commit suicide" as they differentiate into sap-conducting tubes in the veins of developing leaves, in growing stems, and indeed in all developing organs. The death of these cells stimulates further growth and hence further cell death and more auxin production. This research led me to develop a general theory of the aging, death, and regeneration of cells in both plants and animals: cells are regenerated by growth, while the cessation of growth leads to senescence and death.[2]

In India I did research on the physiology of pigeonpeas, pod-bearing shrubs whose flexible branches are used for basket making, much as willows are used in Europe. One of the most successful aspects of my work was the study of regenerative growth, now the basis of a new cropping system involving multiple harvests from the same plants.[3] More recently, I have been developing a way of understanding living nature in terms of inherent memory, described in my books *A New Science of Life* and *The Presence of the Past*. In retrospect, these seemingly disparate activities all look like variations on the theme of the sprouting willow stakes. Likewise, this book is a response to the idea that nature, which we have treated as dead and mechanical, is in fact alive; it is coming to life again before our very eyes.

I studied biology at school and at Cambridge because of my strong interest in plants and animals, an interest encouraged by my father, an herbalist, pharmacist, and amateur microscopist, and accepted by my mother, who helped feed my varied collection of animals and put up with annual invasions of tadpoles and caterpillars. But as I advanced in my studies, I was taught that direct, intuitive experience of plants and animals was emotional and unscientific. According to my teachers, biological organisms were in fact inanimate machines, devoid of any inherent purposes, the product of blind chance and natural selection; indeed the whole of nature is merely an inanimate machinelike system. I had no problem in assimilating this scientific education and through practicals in the laboratory, progressing from dissection to vivisection, acquired the necessary emotional detachment. But there was always a tension; my scientific studies seemed to bear so little relation to my own experience. The problem was summed up for me one day in a corridor in the Biochemistry Department when I saw a wall chart of metabolic pathways, across the top of which someone had written in big blue letters: KNOW THYSELF.

I later came to recognize that the conflict I experienced so intensely was a symptom of a split that runs through our entire civilization. This split is experienced to differing degrees by almost everyone. It is now threatening our very survival.

From the time of our remotest ancestors until the seventeenth century, it was taken for granted that the world of nature was alive. But in the last three centuries, growing numbers of educated people have come to think of nature as lifeless. This has been the central doctrine of orthodox science—the mechanistic theory of nature.

In the official world—the world of work, business, and politics—nature is conceived of as the inanimate source of natural resources, exploitable for economic development. This is the sense of nature that is taken for granted, for example, in *Nature*, a leading international scientific journal. The mechanistic approach has provided us with technological and industrial progress; it has given us better means of fighting diseases; it has helped transform traditional agriculture into agribusiness and animal husbandry into factory farming; and it has given us weapons of unimaginable power. Modern economies are built upon these mechanistic foundations, and everyone is influenced by them.

In our unofficial, private world, nature is most strongly identified with the countryside as opposed to the city, and above all with unspoiled wilderness. Many people have emotional connections with particular places, often places associated with their childhood, or feel an empathy with animals or plants, or are inspired by the beauty of nature, or experience a mystical sense of unity with the natural world. Children are frequently brought up in an animistic atmosphere of fairy tales, talking animals, and magical transformations. The living world is celebrated in poems, songs, and chants, and reflected in works of art. Millions of urban people dream of moving to the country or retiring there, or of having a second home in rural surroundings.

Our private relationship with nature presupposes that nature is alive and usually, at least implicitly, feminine. The approach of the mechanistic scientist, technocrat, economist, or developer is based on the assumption, at least during working hours, that nature is inanimate and neuter. Nothing natural has a life, purpose, or value of its own; natural resources are there to be developed, and their only value is the one placed on them by market forces or official planners.

Another way of looking at this division is in terms of rationalism and Romanticism, established in polar opposition in the late eighteenth century. Then, as now, rationalists were seemingly supported by the successes of science and technology, and Romantics by the undeniable intensity of personal experience. For Romantics, rationalism is unromantic; for rationalists, Romanticism is irrational. We are all heirs to both these traditions, and to the tension between them.

For several generations, Westerners have grown used to living with these internal divisions. A comparable split has now been established in Eastern Europe, Japan, China, India, and to some extent in all the "lesser developed" countries. The missionaries of mechanistic progress have spread their doctrine to all the nations of the world, superimposing it on more traditional, animistic attitudes.

In the first part of this book, I explore the roots of the division between our sense of nature as alive and the theory of nature as dead. This is not merely a matter of historical interest. We are all influenced by mechanistic habits of thought that shape our lives, usually unconsciously. If we are to hold these assumptions up to scrutiny, we need to look at their cultural origins and trace their development. We have to remember that what are now commonplace assumptions were once controversial theories, rooted in peculiar kinds of theology and philosophy, believed only by a handful of European intellectuals. Through the successes of technology, the mechanistic theory of nature is now triumphant on a global scale; it is built into the official orthodoxy of economic progress. It has become a kind of religion. And it has led us to our present crisis.

In the second part, I show how science itself has begun to transcend the mechanistic worldview. The idea that everything is determined in advance and in principle predictable has given way to the ideas of indeterminism, spontaneity, and chaos. The invisible organizing powers of animate nature are once again emerging in the form of fields. The hard, inert atoms of Newtonian physics have dissolved into structures of vibratory activity. The uncreative world machine has turned into a creative, evolutionary cosmos. Even the *laws* of nature may not be eternally fixed; they may be evolving along with nature.

Simple though the idea of living nature may sound, it has profound implications, discussed in the final part of this book. It upsets deep-seated habits of thought; it points toward a new kind of science, a new understanding of religion, and a new relationship between humanity and the rest of the living world. It is in harmony with the idea of the earth as a living organism and with the greening of our political and economic attitudes. We urgently need to find practical ways of reestablishing our conscious sense of connection with living nature. Recognizing the life of nature demands a revolution in the way we live our lives. And we have no time to lose.

PART ONE

HISTORICAL
ROOTS

ONE

MOTHER NATURE AND THE DESECRATION OF THE WORLD

Mother Nature

Like human mothers, nature has always evoked ambivalent emotions. She is beautiful, fertile, nurturing, benevolent, and generous. But she is also wild, destructive, disorderly, chaotic, smothering, death dealing—the Mother in her terrifying form, like Nemesis, Hecate, or Kali.

The idea of nature as a mechanical, inanimate system is in some ways more comforting; it gives a sense that we are in control and gratifyingly confirms our belief that we have risen above primitive, animistic ways of thinking. Mother Nature is less frightening if she can be dismissed as a superstition, a poetic turn of phrase, or a mythic archetype confined to human minds, while the inanimate natural world remains there for us to exploit. Unfortunately the consequences of this way of thinking are themselves terrifying. Nemesis is now operating on a global scale: The climate is changing. We are threatened with droughts, storms, floods, famines, chaos. Ancient fears are returning in new forms.

Although the conquest of nature for the sake of human progress is the official ideology of the modern world, the old intuition of nature as Mother still affects our personal responses and gives emotional force to phrases such as "nature's bounty," "the wisdom of nature," and "unspoiled nature." It also conditions our response to the ecological crisis. We feel uncomfortable when we recognize that we are polluting our own Mother; it is easier to rephrase the problem in terms of "inadequate waste management." But today, with the rise of the green movement, Mother Nature is reasserting herself, whether we like it or not. In particular, the acknowledgment that our planet is a living organism, Gaia, Mother Earth, strikes a responsive chord in millions of people; it reconnects us both with our personal, intuitive experience of nature and with the traditional understanding of nature as alive.

The very words for nature in European languages are feminine—for example *phusis* in Greek, *natura* in Latin, *la nature* in French, *die Natur* in German. The Latin word *natura* literally meant "birth." The Greek word *phusis* came from the root *phu-* whose primary meaning was also connected with birth.[1] Thus our words *physics* and *physical*, like *nature* and *natural*, have their origins in the mothering process.

One of the primary meanings of nature is an inborn character or disposition, as in the phrase *human nature*. This in turn is linked to the idea of nature as an innate impulse or power. On a wider scale, nature is the creative and regulative power operating in the physical world, the immediate cause of all its phenomena. And hence *nature* comes to mean the natural or physical world as a whole. When nature in this sense is personified, she is Mother Nature, an aspect of the Great Mother, the source and sustainer of all life, and the womb to which all life returns.

In archaic mythologies, the Great Mother had many aspects.

She was the original source of the universe and its laws, and the ruler of nature, fate, time, eternity, truth, wisdom, justice, love, birth, and death. She was Mother Earth, Gaia, and also the goddess of the heavens, the mother of the sun, the moon, and all heavenly bodies—like Nut, the Egyptian sky-goddess (Fig. 1.1), or Astarte, the goddess of heaven, queen of the stars. She was Natura, the goddess of Nature. She was the world soul of Platonic cosmology, and she had many other names and images as the mother and matrix and sustaining force of all things.[2]

These feminine associations play an important part in our thinking; our conception of nature is intertwined with ideas about the relations between women and men, between goddesses and gods, and between the feminine and the masculine in general. If we prefer to reject these traditional sexual associations, what are the alternatives to the idea of nature as organic, alive, and motherlike? One is that nature consists of nothing but inanimate matter in motion. But in this case we only deny the mother principle by being unaware of it; the very word for matter is derived from the same root as *mother*—in Latin, the corresponding words are *materia* and *mater*—and (as discussed in Chapter 3), the whole ethos of materialism is permeated with maternal metaphors.

The conception of nature as a machine brings another set of metaphors into play. Many mechanists assume that this way of thinking is uniquely objective, whereas they see the idea of living nature as anthropocentric, nothing but a projection of human ways of thinking onto the inanimate world around us. But surely the machine metaphor is *more* anthropocentric than the organic. The only machines we know of are man-made. Machine making is a uniquely human activity, and a relatively recent one too. The seventeenth- and eighteenth-century conception of God as the designer and creator of the world machine cast him in the image of technological man. And in attempting to see all aspects of nature as machinelike, we project current technologies onto the world around us. Clockwork and hydraulic projections were in

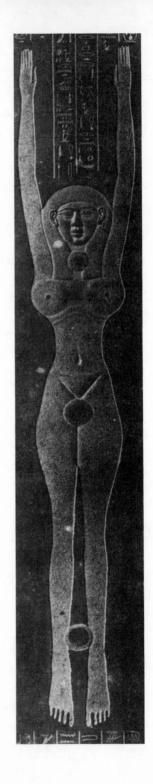

Figure 1.1. The Egyptian sky goddess Nut, portrayed inside a coffin lid (c. sixth century B.C.). Her realm was the vault of heaven, and she gave birth to the sun (represented by discs) each morning and swallowed it at night. (From the British Museum.)

vogue in the seventeenth century, billiard balls and steam engines in the nineteenth century, and computers and information technologies today.

We cannot help thinking in terms of metaphors, analogies, models, and images; they are embedded in our language and in the very structure of our thought. Both animistic and mechanistic thinking are metaphoric. But whereas mythic and animistic thinking depends on organic metaphors drawn from the processes of life, mechanistic thinking depends on metaphors drawn from man-made machinery.

Since the earth is our immediate home, Mother Earth was recognized before the wider domain of Mother Nature was conceived of on a cosmic scale to include the vast expanse of the heavens. The image of the earth as mother is found in traditional cultures all over the world. In the late nineteenth century, this is how a Native American, chief of the Wanapum tribe, explained why he refused to till the ground:

> Shall I take a knife and tear my mother's bosom? Then when I die she will not take me to her bosom to rest. You ask me to dig for stone! Shall I dig under her skin for her bones? Then when I die I cannot enter her body to be born again. You ask me to cut grass and make hay and sell it, and be rich like white men! But how dare I cut off my mother's hair?[3]

The earth was sacred both as the source of life and as the receiver of the dead. She "bringeth all things to birth, reareth them, and receiveth again into her womb," as the Greek poet Aeschylus put it in the fifth century B.C.[4] In many parts of the world, newborn babies were placed on the ground and then picked up again to represent their birth from the womb of the earth. At the same time the ceremony consecrated them to her and ensured that she would protect them.[5] And to this day, even in modern industrial societies, many people still want to be buried in their native land, to return to their earthly womb.

For many millennia, caves played an important part in the religious life of humanity. The earliest-known paintings are found deep within caverns, such as those of Lascaux in southwest France, and they probably played an important part in initiatory journeys undertaken by inhabitants of Europe over twenty thousand years ago. The mystery cults of ancient Greece, such as that celebrated in the cave at Eleusis, continued this ancient tradition. Going into the darkness of the cave was like entering the womb of Mother Earth; emerging again after ritual initiation was like being reborn. And vaults, crypts, and sepulchers are man-made caves in which the bodies of the dead are returned to the womb of the earth.[6]

Even today, caves continue to fascinate millions of people. They are popular tourist attractions. But at the same time, they can be seen as places of pilgrimage to an archaic region of our collective imagination, the underworld, inhabited by the shades of the departed.[7] They are also a gateway into the mineral kingdom and to the material relics of past eras. Erasmus Darwin, Charles Darwin's grandfather, described his journey into the Blue John caverns in Derbyshire, England, in deliberately antiquated terms when he wrote: "I have seen the Goddess of Minerals naked, as she lay in her inmost bower."[8] On this same journey, in 1767, Darwin was deeply impressed by the fossilized shells and bones he found in the caves. "I have been into the bowels of old Mother Earth, and seen wonders and learnt much curious knowledge in the regions of darkness."[9] This experience seems to have sparked off the evolutionary ideas for which he was famous in England until his reputation was eclipsed by that of his grandson.

Mother Earth was seen to be very active. She was thought to exhale the breath of life, which nourished living organisms on her surface. If pressure built up within, she would break wind, causing earthquakes. Fluids flowed within her, and the water came out of her springs like blood. Within her body there were veins, some of which contained liquids, and others solidified fluids like bitumens, metals, and minerals. Her bowels were full of channels, fire chambers, and fissures through which fire and heat were emitted in volcanic exhalations and hot springs. She bore stones and metals within her womb and nurtured them as they grew, like embryos, within her, ripening at their own slow pace.[10]

All over the world, miners traditionally practiced purification rites before entering the womb of the cave or the mine; they were entering a sacred region, a domain that did not rightfully belong to man. The mythologies of mines are replete with fairies, genii, and gnomes, the diminutive guardians of the treasures of the earth. The ores were then taken to the furnace, which speeded up their ripening by heat; the furnaces were like artificial wombs, and the smelter and smith took over the gestatory and formative powers of the Mother. In ancient societies, metalworkers and smiths were at once feared and held in high esteem; their powers were regarded as both sacred and demonic.[11]

With the development of agriculture, Mother Earth gave way to a clearer and more restricted notion of a great goddess of vegetation and harvesting. (In Greece, for example, Gaia was replaced by Demeter.) But women were still very closely associated with the fertility of the soil, and they played a dominant role when agriculture was in its infancy; indeed they may well have invented agriculture.[12] All over the world, metaphors connect women with the ploughed earth, the fertile furrow. In an ancient Hindu text, for example: "This woman is come as a living soil: sow seed in her, ye men!" In the Koran: "Your wives are to you as fields."[13] This same metaphor is implicit in our word *semen*, the Latin for "seed."

Nature was traditionally idealized as benevolent Mother in images of the Golden Age. All was peaceful and fertile; nature gave freely of her bounty; animals grazed contentedly; birds sang pure melodies; flowers were everywhere; and trees bore fruit abundantly. Men and women lived in harmony. There was no disease or strife. The Roman poet Ovid described how in this age the peoples of the world lived in unfortified cities; they enjoyed a leisurely and peaceful existence, had no armor or swords, no use for soldiers.

> And Earth, untroubled,
> Unharried by hoe or ploughshare, brought forth all
> That men had need for, and those men were happy,
> Gathering berries from the mountain sides,
> Cherries, or black caps, and the edible acorns,
> Spring was forever, with a west wind blowing
> Softly across the flowers no man had planted,
> And Earth, unploughed, brought forth rich grain; the field,
> Unfallowed, whitened with wheat, and there were rivers
> Of milk, and rivers of honey, and golden nectar
> Dripped from the dark-green oak trees.
>
> (*Metamorphoses*, Book 1)[14]

Roman poets such as Juvenal coupled this nostalgia with a yearning to escape from the ills of the city, and Virgil wrote of spending his old age "amid familiar streams and holy springs."[15] In the idylls of pastoral poetry, nature has already been subdued: flocks graze peacefully, unharassed by wolves and other predators; the dark forests have been cleared and fertile fields established in their place; the wilderness has given way to orchards and gardens. Nature is calm, kindly, and nurturing, like an ideal wife.

Such visions of the Golden Age have a perennial appeal. Like Ovid, we contrast the peace of earlier times with the strife we experience today; in "primitive" societies we see a harmonious way of living that we have lost—and that these societies themselves are rapidly losing under the influence of our civilization. For millions of modern city dwellers, life is made more tolerable by the prospect of retiring to the country, having a weekend retreat in rural surroundings, or getting away from it all on holiday. On Friday evenings, the roads leading out of the great cities of the Western world are clogged. There is something to be found "in nature" that many of us feel we need.

> Oh there is a blessing in this gentle breeze,
> A visitant that while it fans my cheek
> Doth seem half-conscious of the joy it brings
> From the green fields, and from yon azure sky.
> Whate'er its mission, the soft breeze can come
> To none more grateful than me; escaped
> From the vast city, where I long had pined
> A discontented sojourner.
> (William Wordsworth: *The Prelude*, Book 1, vv. 1–8)

The discontented sojourners in cities have a different sense of nature's power from those whose lives are spent much closer to nature. Storms and droughts, diseases, wild animals, the dangers of the darkness, the forest, and the desert are all too real to those who live outside the relative security of towns or cities. Our fear of wild, untamed nature feeds the desire to subdue her, a desire at least as old as civilization.

The Triumph of the Gods

The ancient image of the Golden Age, usually regarded as a mythic or poetic fantasy, has recently gained a new lease on life as a result of archaeological research in southern Europe and Turkey. The origins of settled agriculture in Europe have now been pushed back to about 7000 B.C. For several thousand years, these early agricultural societies lived in comfortable and usually unfortified settlements, worshiping goddesses, making superb ceramics rather than weapons.[16] But between 4000 and 3500 B.C. this peaceful way of life was shattered by waves of invaders whose warrior gods dethroned the old goddesses, demoting them to wives, daughters, and consorts in the new male-dominated pantheons. Patriarchy and male domination replaced the older, more harmonious social order.[17]

Meanwhile, in the Near East, the old goddess-worshiping agricultural societies likewise came to be dominated by fortified city-states and warring empires. Violent sky-gods became predominant, the vengeful senders of thunderbolts, floods, droughts, famines; the destroyers of cities.[18] The same happened in India, where old relatively peaceful agricultural societies were conquered by invading Aryan warriors, with their sky-gods and horses. The pattern has often been repeated.

From a feminist point of view, this looks like historical evidence that our ills stem from male domination. It also supports the hope that things could be otherwise; a different kind of society actually existed, and could be possible again if we replaced the values of domination and patriarchy with the values of partnership and the Goddess.[19]

However, the Neolithic revolution gave rise to two very different kinds of society: settled agriculturalists and nomadic or semi-nomadic pastoralists. Even if we think of the first few millennia of agriculture as a Golden Age for the settled peoples, the pastoralists led a far less comfortable existence on the fringes of the desert and in the steppes. The pastoralists, who were essentially domesticated hunters, generally worshiped sky-gods, were patriarchal, and valued masculine force, bravery, and fortitude. Early agricultural societies coexisted with these nomadic groups and in many cases had a symbiotic relationship with them. But somehow the pastoralists triumphed over the more peaceful, feminine cultures of the settled people. They could easily transform themselves into warriors, hunting and killing people instead of wild animals. They could dominate and enslave other people, just as they controlled herds of animals.

The triumph of the warriors was reflected in new myths. To start with, the primal Mother was the source of all things. She was the Virgin Mother; she needed no god in order to conceive, and all the gods were descended from her. For example, according to one of the early Greek creation myths, first of all Mother Earth (Gaia) emerged from chaos. She then gave birth to Uranus, the sky-god, as she slept. He was both her son and lover; gazing down at her from the mountains, he showered fertile rain upon her secret clefts, and she bore grass, flowers, and trees, and brought forth the birds and the beasts.[20] The great shrine and oracle of Gaia was at Delphi, the center of the cosmos. But then her great-grandson Apollo slew the great python at Delphi and usurped Gaia's shrine. Nevertheless, Gaia remained the source of prophetic power, and her priestess, the pythoness, continued to prophesy underneath Apollo's temple.

In the earliest Babylonian creation stories, the primal goddess Tiamat was the formless void, the deep, the dark womb from whom the universe was born; she brought forth the world by herself. The god Marduk was originally her son. But then Marduk became the creator god, slaying Tiamat, now pictured as the dragon of chaos. He smashed her skull, split her body like an oyster, and the obedient winds whisked her blood away. By dividing her in two, he created the firmament of heaven and foundation of earth.

In the first chapter of the book of Genesis, the primal Mother is again the formless void, the dark, the watery abyss. Unlike Marduk, God did not fight with her: "The earth was a vast waste, darkness covered the deep, and the spirit of God hovered over the surface of the water" (Genesis 1, 2). But like Marduk, he first created by dividing—the light from dark, day from night, the waters above from the waters below, the heavens from the earth, and the dry land from the seas. Both the earth and the seas retained the fertile, creative powers of the Mother. When it came to the creation of plants, God evoked them, but he did not make them himself. They were formed and brought forth by Mother Earth:

God said, "Let the earth produce growing things; let there be on earth plants that bear seed, and trees bearing fruit each with its own kind of seed." So it was; the earth produced growing things.

(Genesis 1:11–12)

Likewise he called forth land animals from the earth, and sea creatures and birds from the waters. In theological terminology, this was not a direct but a "mediate" mode of creation.[21]

Although the Judeo-Christian tradition has always emphasized the supremacy of the male God, Mother Earth retained some of her old autonomy for many centuries. The Jews were forbidden to worship the old goddesses, but nevertheless the Holy Land remained both sacred and female, and Jerusalem herself was the bride of God. Throughout the Middle Ages, Christians continued to regard nature as animate and motherlike.

The complete supremacy of the Father was not established until the Protestant Reformation in the sixteenth century, with the suppression of the Holy Mother's cult and the desacralization of the natural world. This process was taken to its ultimate conclusion in the seventeenth century, when nature became nothing but inanimate matter in motion, created by God and mechanically obedient to his eternal laws. Nature was no longer acknowledged as Mother, and no longer considered alive. She became the world-machine, and God the all-powerful engineer.

Ironically, the idea that nature functioned mechanically and automatically rendered God increasingly superfluous, and by the late eighteenth century, he was fading away from the scientific worldview. With the subsequent growth of atheism, nature alone was seen as the source of everything. Indeed, in order to account for the creativity of the evolutionary process, she had to be credited more and more freedom and creativity. For the modern materialist, nature or matter is the source of all things; all life emerges from her, and to her all life returns. She may no longer be venerated, but she has taken on some of the most fundamental properties of the Great Mother. Just as the gods were once thought to be descended from the primal Mother, so in the eyes of the modern materialist are they descended from matter. From the blind material processes of evolution, human minds emerged; and from human minds the gods arose by a process of psychological projection.

The Loss of the Sacred World

Nowadays we live in a desacralized world. Of course certain seasonal festivals, such as Easter and Yom Kippur, still have a religious significance for the faithful, and so do certain places, such as Lourdes and Mecca, or certain animals and plants, such as cows and pipal trees for Hindus. But there is nothing in the scientific worldview to support such ideas of sacredness. They are survivals from earlier epochs.

From the conventional modern point of view, our ancestors, like primitive peoples all over the world, were unable to see nature as it is—an inanimate, purposeless, physical system— because they projected onto it their own hopes, fears, and fantasies. They were victims of the pathetic fallacy, attributing to inanimate objects the characteristics of animate creatures. They filled the world around them with goddesses, gods, spirits, souls, and nonhuman powers; they endowed particular places and times with mystical significance because of primitive, animistic, and superstitious habits of thought. These processes were encouraged and exploited by shamans, priests, and magicians whose power was enhanced by such ignorance and superstition. But thanks to the advances of science and the growth of rational understanding, we now know that nature cannot be influenced by spells and incantations, by rituals and mumbo jumbo. Rather, it is governed by impersonal laws that operate uniformly at all times and in all places. Many things also happen by chance, but such random events have nothing to do with the activity of spirits or divine interventions. We can have no power over nature by magic or through mystical forces, and we cannot hope for miracles. What we *can* do is gain ever-increasing mastery through science and technology.

These familiar opinions—the doctrines of secular humanism—are closely related to the mechanistic theory of nature that has dominated scientific thinking since the seventeenth century. However, the process of undermining the sacredness of nature actually began much earlier. It was already well advanced in northern Europe as a result of the Protestant Reformation in the sixteenth century. This religious revolution not only helped pave the way for the development of modern science but also provided a favorable environment for the growth of technology and the acceleration of economic development.[22] Traditional religious and symbolic values attached to particular places, plants and animals were replaced by monetary values. We still see this conflict of attitudes as native peoples struggle, generally unsuccessfully, to save their sacred places from being taken over for mining and other forms of economic development, and closer to home in the recurrent conflicts between conservationists and developers.

The Reformation led to a contraction of the spiritual realm, a withdrawal of the spirit from the operations of nature. The realm of the spirit was concentrated within human beings; the rest of the natural world was just the background against which the human spiritual drama was played out. Modern secular humanism has of course abandoned belief in an afterlife, but most of its essential features are derived from the Protestant tradition, including its attitude to nature.

Despite all this, a vague sense of the sacredness of nature, an unarticulated nostalgia, persists in many of us. The widespread desire to get back to nature, the need to find inspiration in the countryside or untamed wilderness, stems from this residual—now fragmented and disjointed—sense of the sacred. In traditional societies there is a collective recognition of sacred places and times, and a mythic framework that gives them their significance. But modern secular life has left such beliefs behind; deprived of the possibility of expression in religious forms, such feelings are most intensely experienced in solitude. They are "merely subjective" in the sense that they correspond to nothing in the inanimate physical world of scientific theory; nor can they be recognized collectively through appropriate ceremonies and observances. They can be categorized as "poetic," "romantic," "aesthetic," or "mystical." But as such they can be part only of our private lives.

The Righteousness of Desecration

In sacred places, the spiritual and the physical are experienced together. Sacred places are openings between the heavens and the earth, or between the surface of the earth and the underworld; they are places where different planes or levels of experience cross. In ancient Palestine, as in many other parts of the world, certain megaliths or standing stones were gateways of this kind. Such a stone, in wild and desolate surroundings, was venerated by the Jews at Bethel. At this place Jacob was said to have had a dream of a ladder reaching up to heaven, with angels ascending and descending on it. From above the ladder, God spoke to him, saying: "The land on which you are lying I shall give to you and your descendants" (Genesis 28: 13). When Jacob woke from his sleep, he said:

> "Truly the Lord is in this place, and I did not know it."
> He was awestruck, and said, "How awesome is this place! This is none other than the house of God; it is the gateway to heaven." Early in the morning, when Jacob awoke, he took the stone on which his head had rested, and set it up as a sacred pillar, pouring oil over it.
>
> (Genesis 28:17–18)

It is impossible to know whether this was a sacred place because of Jacob's experience there, whether he had his vision there because it was already a special place of power, or whether this story grew up to account for the fact that worship and sacrifice had been practiced there since time immemorial.

When the Jewish people entered the Promised Land as pastoralist warriors, it was already inhabited by Canaanites, Philistines, and other peoples. As they settled there and adopted an agricultural way of life, local agricultural festivals were assimilated into their religion, as were many of the ancient places of power such as holy wells (for example, at Beersheba—Genesis 26:24) and sacred oak and terebinth trees (for example, the terebinth of Moreh—Genesis 12:6–9). For many generations, they worshiped at the old "high places" and groves sacred to the Queen of Heaven. These sanctuaries were equipped with stone pillars, altars for animal sacrifice, and tree stumps known as *asherahs*, the name of the old goddess.[23]

The religion of the Jews resembled many others, both pastoralist and agriculturalist, in the recognition of sacred places and times, and in the killing of sacrificial animals. But one of the ways it differed was its insistence on the uniqueness of its God; another was through the prohibition on the making and worshiping of images. God was to be known through the natural world, his own creation, rather than through idols made by man. These features of Judaism were inherited by Christianity and Islam, and have had profound historical effects.

Much of the history of the Jews recorded in the Old Testament concerns their conflicts with the indigenous peoples of Palestine. The prophets, reminding the people of Israel of their nomadic heritage, rejected the indigenous goddesses and gods, and continually denounced the tendency to adopt the religious practices of the surrounding peoples. Nevertheless, for many centuries, worship continued at the old high places and sacred groves, and the cult of the sacred snake and the worship of goddesses persisted.

When King David conquered Jerusalem, the hilltop site for a temple was revealed to him, and it was duly built by his son

Solomon. At first, in spite of its magnificence, this was just one of the many places where sacrifices were offered. But the idea began to develop that the one God should be worshiped in only one place, the temple. Several attempts were made by kings of Jerusalem to suppress all other places of worship, to purify the cult of Jahweh and to center the religion in the city. One wave of desecration took place under King Hezekiah, who cut down the sacred groves, defiled the high places, destroyed the images, and "broke up the bronze serpent that Moses had made, for up to that time the Israelites had been in the habit of burning sacrifices to it" (2 Kings 18:4).

Nevertheless, the old practices continued, and some eighty years later, around 622 B.C., King Josiah took yet more violent steps. He desecrated the hill shrines, slaughtered the priests on their altars, burned the sacred groves, and ground the sacred stones to dust, including the stone at Bethel (2 Kings 23:14–15). He defiled the shrines established in Jerusalem by Solomon to the goddess Ashtoreth and other goddesses and gods. But scarcely had a generation passed when Jerusalem itself was conquered, the temple desecrated, and many of the Jews led captive into Babylon (2 Kings 25).

Nineteenth-century travelers in the Holy Land frequently described hilltop shrines, often found in groves of oaks or terebinths, with small domed buildings or white painted stones. These were venerated by the local Muslim peasants as places where saints were said to have stayed or where they had been buried. The trees themselves were regarded as sacred, and their fallen branches could not be collected for firewood. Thus the worship at the high places and sacred groves, which pious Hebrew kings forbade and prophets thundered against thousands of years ago, persisted, apparently in the same places, right into modern times.[24]

The pre-Christian religions of Europe, like the pre-Jewish religions of Palestine, were polytheistic; they involved a variety of seasonal ceremonies and rituals, and they recognized many sacred places, including trees, wells, groves, rocks, standing stones, mountains, and rivers. During the conversion of Europe from the worship of the old gods and goddesses, many of the traditional sacred places and seasonal ceremonies were retained in a Christianized form (Fig. 1.2). This incorporation of archaic religious elements into the Christian religion is still obvious in Roman Catholic and Orthodox countries. Think, for example, of the holy wells and springs in Ireland or the sacred mountain Croagh Patrick, a major center of pilgrimage.

In some cases Christianity was seen as a development or culmination of the old religion; in the Celtic church in Ireland and Britain, for example, many of the early saints seem to have achieved a remarkable harmony between the druidic past and the new religion. To the old sacred places were added new ones connected with these saints: places where they had seen visions, where they had lived and died, and where their relics were enshrined.[25] In other cases, the old religions were assimilated as a matter of deliberate papal policy. Here are some of the instructions of Pope Gregory the Great to St. Augustine of Canterbury, sent to evangelize the English at the end of the sixth century:

> Because the English have been used to slaughtering many oxen as sacrifices to devils, some solemnity must be put in place of this. On the feast days or birthdays of the holy martyrs whose relics are there deposited, close by those churches which were once pagan temples, they may build themselves huts of the boughs of trees and hold a religious feast. Instead of offering beasts to the devil, they shall kill cattle and in eating them praise God.[26]

There is no doubt that the spread of the cult of Mary involved the assimilation of various elements of pre-Christian goddess worship.[27] Indeed the fifth-century council of the church at which she was proclaimed Mother of God took place at Ephesus, an ancient center of goddess worship, only a few decades

Figure 1.2. A Christianized megalith in Brittany. The menhir of Champ-Dolent, near Dol (Ille-et-Vilaine). (From Nodier and Taylor, 1845.)

after the temple of Artemis was suppressed. Mary is Queen of Heaven, a title inherited from Astarte-Ashtoreth, the aspect symbolized by her blue, star-spangled cloak; she is lunar, like Artemis, and is often depicted standing on a crescent moon; she is the Star of the Sea, with many sanctuaries all around the shores of the Mediterranean; and as Virgin Mother of God, she is heir to the ancient tradition of the primal Mother. She also took on aspects of the Earth Mother, not least through her shrines in caves, grottoes, and crypts, and as the protectress of many holy wells. And like the Great Mother who gave life and took it back again, she is present at death. A plea for her protection at death is the final line of the Hail Mary, the *Ave Maria*: "Holy Mary, Mother of God, pray for us sinners now and at the hour of our death."

The Protestant reformers were trying to establish a purified form of Christianity, rejecting the corruptions and abuses of the Roman church. Personal faith and repentance were what mattered; ritual observances, seasonal festivals, pilgrimages, devotion to the Holy Mother, and the cults of saints and angels were all denounced as pagan superstitions. As John Calvin justly observed, "Nuns came in place of vestal virgins; the church of All Saints to succeed the Pantheon; against ceremonies were set ceremonies not much unlike."[28] Sharing in the contemporary humanist respect for scholarship and for fidelity to original sources, the reformers used the Bible as their source of authority, rejecting many of the subsequent doctrines and traditions of the church.[29] Obviously they found no justification in the Bible for practices which had grown up since it was written. The power of the papacy, the doctrine of purgatory, the cults of Mary and the saints, the use of images, and the reverence for the sacred places of Europe—they denounced all these as pagan and launched an orgy of destruction.

Images of the Holy Mother and of saints and angels were broken and burned; stained-glass windows were smashed; holy wells and wayside shrines were defiled; the tombs of saints broken open and their relics scattered; pilgrimages suppressed; many of the customary rituals and ceremonies abolished; monasteries and convents plundered and ruined. In some cases this destruction was the work of fanatics.

> No-one who sees the iconoclasts raging thus against wood and stone would doubt that there is a spirit hidden in them which is death-dealing, not life-giving, and which at the first opportunity will also kill men. (Martin Luther, 1525)[30]

In other cases the destruction was carried out as a systematic political policy. In England, from 1536 to 1540, King Henry VIII's commissioners dissolved the monasteries, appropriating their treasures and their lands, and ejecting the monks and nuns. They also turned their attention to great shrines such as those of St. Thomas at Canterbury and St. Hugh at Lincoln. The king's warrant for their visit to Lincoln was typical in its purposes: "to bring our loving subjects to the right knowledge of the truth, taking away all occasions of idolatry and superstition," and, at least as importantly, "to see the said relics, jewels, and plate, safely and surely to be conveyed to our Tower of London."[31]
The Protestants were trying to bring about an irreversible change in attitude, eradicating the traditional idea that spiritual power pervades the natural world and is particularly present in sacred places and in spiritually charged material objects. They wanted to purify religion, and this purification involved the disenchantment of the world.[32] All traces of magic, holiness, and spiritual power were to be removed from the realm of nature; the spiritual realm would be confined to human beings. Even the elements of the Christian sacraments were denied their spiritual power. The reformers argued that believing in the real presence of Christ in the consecrated bread and wine of the mass was analogous to believing in the real presence of saints in their con-

secrated images or in the power of holy water, sacred relics, or hallowed ground. All were superstitious and idolatrous.[33]

The material world was governed by God's laws and incapable of responding to human ceremonies, invocations, or rituals; it was spiritually neutral or indifferent and could not transmit any spiritual power in or of itself. To believe otherwise was to fall into idolatry, transferring God's glory to his creation. There was to be no attempt through religious means to change the way the natural world operated; it should be accepted as an expression of God's will. In the influential view of Calvin, God had predestined all events from the beginning of time. He could, if he chose, break into the material realm to bring about miracles to communicate with humans. He had done so in the early church in order to spread the Gospel among heathens, but those days were now over, and there was normally no intrusion of the divine spiritual sphere into the material.[34]

The Reformation thus prepared the ground for the mechanistic revolution in science in the following century. Nature was already disenchanted and the material world separated from the life of the spirit; the idea that the universe was merely a vast machine fitted well with this kind of theology, and so did the constriction of the realm of the soul to a small region of the human brain. The domains of science and religion could now be separated: science taking the whole of nature for its province, including the human body; religion, the moral and spiritual aspects of the human soul.

Although Kings Hezekiah and Josiah set precedents for the Protestant desecrations, their aims were very different. They were not trying to stamp out the idea that places and times could be sacred or denying the importance of sacrifices and festivals; they were trying to centralize the Jewish religion in the city. They wanted to enhance the holiness of Jerusalem in general and the temple in particular. Nor were the desecrations of old sacred places by Roman Catholic missionaries an attack on the sanctity of the earth; they wanted sacred places to be Christian rather than pagan and often took over the old ones.

The Protestant iconoclasts had a different goal: not the substitution of one kind of sacred place for another but the abolition of all sacred places. At best, everywhere was sacred; at worst, nowhere. The worst view predominated. And although the world was not desacralized through the fervor of Protestant faith alone, the Reformation helped unleash forces that have been accelerating this process ever since.

The Rising Power of Mammon

Some of the Protestant iconoclasts realized that the destruction of external idols was not enough; more would always spring up within. Calvin saw this as a fundamental defect of the human mind: "Surely, just as waters boil up from a vast, full spring, so does an immense crowd of gods flow forth from the human mind."[35] The battle against the idols could never be won just by destroying images.

The most powerful of the idols in the desacralized world was Mammon. In the New Testament, he was the personification of riches; by the Middle Ages, he had become the demon of commercial greed. The greatest of the Puritan poets, John Milton, depicted him as a fallen angel:

> . . . even in heaven his looks and thoughts
> Were always downward bent, admiring more
> The riches of Heaven's pavement, trodden Gold,
> Than aught divine or holy else enjoyed
> In vision beatific: by him first
> Men also, and by his suggestion taught,
> Ransacked the Centre, and with impious hands
> Rifled the bowels of their mother Earth
> For Treasures better hid. Soon had his crew

> Opened into the Hill a spacious wound
> And digged out ribs of Gold.
>
> (*Paradise Lost*, Book 1, 680–90)

Far older than the Christian conception of Mammon is the Sumero-Babylonian goddess Mammetun, the Mother of Destinies. Her name may stem from the same linguistic root as our words *mama*, *mammary*, *mammal*, and *mother*. And Mammon could be a masculine form of the name of the archaic goddess whose generous breasts were the source of plenty.[36] The appropriation of her gifts by men was diabolical, and Mammon was a male demon.

In India, wealth is still thought to flow from a goddess, Lakshmi, who is often depicted pouring out streams of gold coins from her two inexhaustible vases of plenty. In ancient Rome, money was minted in the temple of Juno Moneta, the Great Mother in her aspect of adviser and admonisher.[37] She is the source of our words *money* and *monetary*.

Money has many metaphorical aspects. Gold coins were like little images of the sun. But modern money is more alive, filled with a breathlike spirit, subject to inflation and deflation. Money is also currency, and its flow is what animates the economy; like blood, it circulates. Monetary assets are "liquid." Like a milk-giving breast or udder, the economy works on the basis of supply and demand; it supplies the demands of consumers. And, like a woman, its behavior is cyclic. Money is a human creation, and so is the economy that generates it, but it has taken on a life of its own. Economic forces rather than natural forces have come to dominate our lives, and the ruling power of our world is Mammon.

The Final Desecration

The rationalistic spirit in which the Protestant reformers attacked the practices of traditional religion was not brought to bear on their own beliefs. These depended on faith and on the authority of scripture. But once the forces of skepticism and iconoclasm were released, there was no stopping them. Secular humanism takes the Reformation to its ultimate extreme, turning the Protestant critique onto Protestant faith. In this second revolution, devotion to the words of the Bible in itself becomes a form of idolatry. What reason is there to accept its authority? As for God, why should he not be like other gods, a phantom of the human mind? Those whose religion is founded on a protest against other people's unreasoning faith have been ill equipped to defend an unreasoning faith of their own.

This ultimate reformation, the protest against Protestantism, leaves man as the source of all goddesses and gods, the master of desacralized nature, the only conscious rational being in an inanimate world. For the secular humanist, nothing is sacred except human life. Indeed humanism can easily become a religion itself, glorifying man and his wondrous works. But the spirit of negation will never be far away: why should man be sacred? He is just another species thrown up by the blind forces of evolution and no doubt doomed to extinction like the dinosaurs. In the end, nothing at all is sacred.

The results are disastrous. The desecration of the world now seems appallingly destructive, even in the eyes of many humanists. We need to recover a sense of the sacred. Here is a sign of the times, a report that appeared in *Nature* on February 1, 1990, under the heading "Global Change":

> Astronomer Carl Sagan and 22 other well-known researchers chose Moscow as the unlikely venue for an appeal to world religious leaders to join scientists in protecting the global environment. The appeal came at a recent conference on the environment and economic development which attracted over a thousand religious, political and scientific leaders from 83 nations.

Ironically, Sagan travelled to the officially atheist Soviet Union to announce "a religious as well as a scientific dimension" to the problems of global change. Even more remarkable, the conference was sponsored by both the USSR Academy of Sciences and the Russian Orthodox Church.

The appeal states that "efforts to safeguard and cherish the environment need to be infused with a vision of the sacred." Among those who have given their backing are physicist Hans Bethe, biologist Stephen Jay Gould and former MIT President Jerome Weisner.

The appeal certainly reached a global audience. It, and other parts of the five-day conference, were the first ever to be televised with satellite time provided jointly by the East and West communications networks . . . and reached an estimated 2,000 million people in 129 countries. Later at the conference, more than one hundred religious leaders joined to hail the scientists' appeal as "a unique moment and opportunity in the relationship of science and religion."[38]

TWO

THE CONQUEST OF NATURE AND THE SCIENTIFIC PRIESTHOOD

Man's Dominion over Nature

After creating the first man and woman, "God blessed them and said to them, 'Be fruitful, and increase, fill the earth, and subdue it: and have dominion over the fish in the sea, the birds of the air, and over every living thing that moves on the earth' " (Genesis 1:28). This passage, and other similar biblical texts, are often taken to lie at the root of the environmental destruction wrought by modern industrial civilization.[1] But this view is too simple; the problem goes much deeper.

The ancient Greeks, for example, had a view of nature that was, if anything, even more anthropocentric than that of the Jews. In his *Politics*, Aristotle gave the following summary of the order of nature:

Property, in the sense of a bare livelihood, seems to be given by nature herself to all, both when they are first born, and when they are grown up. . . . We may infer that after the birth of animals, plants exist for their sake, and that the other animals exist for the sake of man, the tame for use and food, and the wild, if not all, at least the greater part of them, for food, and for the provision of clothing and various instruments. And so, in one point of view, the art of war is a natural art of acquisition, for the art of acquisition includes hunting, an art which we ought to practise against wild beasts, and against men who, though intended by nature to be governed, will not submit; for war of such a kind is naturally just.[2]

The acquisition of property, including slaves, is justified by making explicit the connection between hunting and war, one of the recurrent ways dominion over nature is closely associated with dominion over other people.

Though Paleolithic societies of hunter-gatherers seem to have lived in greater harmony with nature than agricultural societies or urban civilizations, they still appear to have wrought major changes in their environments. In Southeast Asia, *Homo erectus* may have hunted certain primate species to extinction and changed forever the habitats and populations of species like the orangutan and panda.[3] In Europe and the Americas, humans may have been responsible for the extinction of many species of mammals some ten thousand years ago, such as the giant armadillo in South America, mammoths in northern Europe, and the pygmy hippopotamus on Cyprus, either by excessive hunting or by destroying their environment.[4] A number of large-scale ecological changes in prehistoric times seem to have been due to human activities, including the deliberate burning of great tracts of forest and grassland. Much of the world's desertification may have been aggravated by the activities of prehistoric man.

The technological impulse and the urge to transform our surroundings, to rework nature into forms that reflect human cultures and myths, seem as native to human beings as language and the use of tools and fire. We need to remember too that all living organisms affect their environment to varying degrees: plants in general are responsible for the oxygen in the atmosphere; forests affect the climate; trees shade the ground beneath them, suppressing the growth of other plants; animals compete for food, and some bring about dramatic ecological changes, like beavers with their dams or swarms of locusts with their all-consuming appetites.

Cultures differ in the strength of their impulse toward mastery and in the countervailing sense of kinship with the natural world. But the whole of human history—ever since fire was first tamed, tools first made, metals first used, animals and plants first domesticated, cities first built—has involved man's domination of nature to varying degrees. What is unique in the modern world is not the fact of human power itself nor the sense of the uniqueness of humanity but the vast increase in human power. Mythological, theological, and philosophical justifications for human power over the natural world are not a unique feature of modern civilization or of the Judeo-Christian tradition; they are found everywhere. So are anthropocentric views of man's relationship to nature. Even the heroic imagery of man's conquest of nature, so important in the modern ideology of progress, has ancient precedents. The archetype of the hero conquering wild nature appears in the Babylonian myth of Marduk triumphing over Tiamat, the monster of the deep; the Egyptian god Horus over the hippopotamus; Perseus over the Gorgon; Apollo over the python; and St. George over the dragon.

Thus the recent increase in technological power finds no simple explanation in a specifically Judeo-Christian belief in man's right to subdue the earth and have dominion over other living creatures. As a matter of fact, the Jews neither built the most technically advanced civilization of the ancient world nor stood out above all others in their ambition to dominate nature. In this sense, far more was achieved by the Egyptians, Sumerians, Babylonians, Persians, Greeks, and Romans. Nor have most Christian cultures been noted for their technical achievements. Ethiopia, for instance, has been Christian for many centuries longer than Western Europe but has not dazzled all Africa with its scientific or technical brilliance. Nor did the civilizations of Byzantium and medieval Europe, for all their artistic and technical accomplishments, have much more power over nature than the contemporary civilizations of India or China with their very different religious and philosophical systems.

As we have seen, the relatively recent acceleration in technological mastery is rooted in the scientific revolution of the seventeenth century, which itself grew out of the ferment of the Renaissance and the Reformation. What made the difference was a vast inflation of the ambition to dominate and control nature, a way of treating the natural world as if it had no inherent value or life of its own, and an overthrow of traditional restraints on human knowledge and power.

Not far in the background was the vivid image of the European conquest of America. Gold and fabulous riches were there for the taking. Native peoples were plundered, raped, massacred, infected, enslaved, dispossessed. Their sacred places were desecrated, and their sense of spiritual connection with the land rejected as pagan superstition. The spirit in which this enterprise began was clearly described by Cortés himself. In a letter to the Spanish king, he wrote that his fellow conquerors of Mexico were "not very pleased with [the new rules imposed by Spain], in particular those that bind them to strike root in the land; for all, or most, of them intend to deal with these lands as they did with the Islands first populated, namely to exhaust them, to destroy them, then to leave them."[5]

The success of the conquest of America depended not only on the Europeans' bravery, greed, and sense of religious superiority but also on their more powerful technology, not least their firearms. The new vision of conquering nature developed in the context of the violent enlargement of European dominion over the Americas and ultimately over most of the face of the earth.

Dreams of Power: The Faustian Bargain

One of the most surprising features of the scientific revolution is the way it occurred in a climate of thought permeated by alchemy, magic, mysticism, and a widespread fear of witchcraft.[6] From the beginning of the sixteenth century, there was a growing interest in magical powers and complex systems of sympathetic magic, including a revival of the Hermetic tradition, believed to represent the secret magical teachings of ancient Egypt.[7] The figure who epitomized the quest for superhuman power was Dr. Faust. The first Faust book was published in Germany in 1587, exactly a century before Newton's *Principia*. Faust predated the birth of mechanistic science; he was a magician, but he embodied the desire for unlimited knowledge and power that played such an important part in the mechanistic revolution and remains central to the spirit of science to this day.

Dr. Faust has been the subject of dozens of plays, poems, and novels, and his changing fortunes have reflected the changing spirit of the times.[8] By the early nineteenth century, Faust could no longer be condemned for wanting limitless knowledge and power; in the progressive spirit of the age, such a striving was now thought to be good, not evil. In Goethe's *Faust* (1808) the terms of the contract with the devil were changed accordingly. Faust was not to be carried off to hell at the end of a fixed period but only if he tired of his restless quest, if he ceased to be unsatisfied. Eventually he succumbs to a moment of contentment, imagining that he could say to the fleeting moment, "Stay yet awhile, thou art so fair." But after a brief struggle with the devils who come to claim him, he is saved and carried up to heaven in true baroque style.

In Mary Shelley's *Frankenstein: Or the Modern Prometheus* (1818), Faust takes on a more modern incarnation. Like Faust, Frankenstein is driven by the desire for godlike power. He wants to create life. But the punishment for his pride and presumption is no longer meted out by devils, nor can he be saved by angels.

He is destroyed by the monster of his own creation. The spirit of Frankenstein lives on today, not only in horror movies and in the fantasies of genetic engineers. We have created many monsters that threaten to destroy us, not least our nuclear weapons. The most powerful of all human creations are hydrogen bombs, transmutational devices worthy of their alchemical ancestry, based on a marriage of sun and earth. The sunlike energy released by the fusion of atoms of the lightest element, hydrogen, is detonated by the fission of one of the heaviest, plutonium, named after the god of the underworld.

The Faustian and Frankensteinian aspects of scientific enterprise have never been far below the surface of our consciousness. Sometimes they become explicit. For example, in recent years in Britain, government ministers have argued in favor of nuclear power and nuclear weapons on the grounds that they represent a "Faustian bargain" from which we cannot turn back. Perhaps so. Perhaps, indeed, the Faustian bargain is fundamental to our entire scientific, technological, and industrial system.

Francis Bacon and the Scientific Priesthood

The greatest prophet of the conquest of nature was Francis Bacon. His aim was "to endeavour to establish the power and dominion of the human race itself over the universe." He was well aware of the traditional prohibitions against inordinate ambition, the popular fear of witchcraft, and the damnation of Dr. Faustus. He needed to banish the fear, guilt, and sense of evil traditionally associated with this desire for unlimited power.

Bacon was a lawyer by training and profession, and his abilities and ambition enabled him to rise to the position of Lord Chancellor, the highest legal official in England. In advancing his vision of the mastery of nature, Bacon had to refute the arguments of those who believed that such ambitions were satanic. This was a difficult task, and to this end he used his extraordinary gifts of argument.

He equated the dominion over nature with Adam's naming of the animals (Genesis 2:19–20) in which woman had no part, since it took place before the creation of Eve. He could thus make the technological mastery of nature appear to be a recovery of power given by God to man, rather than something new. At the same time he drew an ingenious distinction between the innocent knowledge of nature, which was beyond good and evil, and morality, in which man should submit to God's commands. In so doing, he established the distinction between facts and values that is so familiar today:

> It was not that pure and uncorrupted natural knowledge whereby Adam gave names to the creatures according to their propriety, which gave occasion to the fall. It was the ambitious and proud desire of moral knowledge to judge of good and evil, to the end that man may revolt from God and give laws to himself, which was the form and manner of the temptation.[9]

Did Bacon sincerely believe in this division between "natural knowledge" and "moral knowledge"? Or was this just a clever lawyer's argument? We cannot be sure, but he certainly seemed confident that we would use our knowledge of nature wisely and well: "Only let the human race recover that right over nature which belongs to it by divine bequest; the exercise thereof will be governed by sound reason and true religion."[10] In retrospect, we can see that he was wrong. Think, for example, of the current devastation of the Amazon forests, made possible by technology and a Baconian faith in man's right to master nature. Sound reason and true religion are nowhere in evidence, and nothing could be further from the innocent exercise of man's God-given right to name the living creatures. Countless species are being killed off unnamed and unknown.

Bacon also removed the power of classical myths, interpreting them as parables instructing man to develop a wholly rational understanding of the natural world.[11] In his *Wisdom of the Ancients* (1609) he took Proteus, for example, as a personification of matter, and his power to change shape represented the capacity of matter "to turn and transform itself into strange shapes" when force is brought to bear on it. Pan represented "the universal frame of things, or Nature." His hairy body corresponded to the rays emitted by objects, and his office was best explained by his role as the god of hunters:

> For every natural action, every motion and process of nature, is nothing else than a hunt. For the arts and sciences hunt after their works, human counsels hunt after their ends, and all things in nature hunt either after their food, which is like hunting for prey, or after their pleasures, which is like hunting for recreation.

The key to this new era of power over nature was organized research. In his *New Atlantis* (1624), Bacon described a technocratic utopia, in which a scientific priesthood made decisions for the good of the state as a whole and also decided which secrets of nature were to be kept secret. Their prototypic research institute, Salomon's House, contained a series of laboratories and artificial environments in which nature could be modeled with a view to controlling her. With astonishing insight, Bacon predicted what institutional science could do. Artificial tempests, for example, could be created for study by using "engines for multiplying and enforcing of winds." New forms of animal and plant life were to be created and existing ones manipulated by experiment; the New Atlantis had parks and enclosures where birds and beasts were maintained for experimental purposes:

> By art likewise we make them greater or taller than their kind is, and contrariwise dwarf them and stay their growth; we make them more fruitful and bearing than

their kind is, and contrariwise barren and not generative. Also we make them differ in colour, shape, activity, many ways.

And of course animals were also maintained for vivisection and for medical research. "We also try all poisons and other medicines upon them."

The general purpose of this foundation was "the knowledge of causes, and secret motions of things; and the enlarging of human empire, to the effecting of all things possible." Bacon's visionary project was a direct inspiration for the founding of the Royal Society of London some forty years later and ultimately for a host of other scientific academies and research institutes. He is justly recognized as a founding father of modern science.

The idea of man's conquest of nature is inseparable from sexual imagery, as feminist scholars have made very clear.[12] The writings of Francis Bacon provide a number of striking examples. Using metaphors derived from contemporary techniques of interrogation and torture of witches, he proclaimed that nature "exhibits herself more clearly under the trials and vexations of art [mechanical devices] than when left to herself."[13] In the inquisition of truth, nature's secret "holes and corners" were to be entered and penetrated. Nature was to be "bound into service" and made a "slave" and "put in constraint." She would be "dissected," and by the mechanical arts and the hand of man, she could be "forced out of her natural state and squeezed and moulded," so that "human knowledge and human power meet as one." He advised the new class of natural philosophers to follow the model of miners and smiths in their interrogation and alteration of nature, "the one searching into the bowels of nature, the other shaping nature as on an anvil."[14] And he wrote of the new science as a "masculine birth" that will issue in a "blessed race of Heroes and Supermen."[15]

Many of the early Fellows of the Royal Society followed Bacon in his use of the epithet "masculine" for privileged and productive knowledge, and continued to speak of capturing, dominating, and subduing nature. Some went further still. Robert Boyle, for instance, decried "the veneration men commonly have for what they call nature," because this "obstructed and confined the empire of man over the inferior creatures." He proposed that "instead of using the word nature, taken either for a goddess, or a kind of semi-deity, we wholly reject, or very seldom employ it."[16] By the end of the seventeenth century, in the eyes of science, nature had ceased to be feminine at all; she simply became inanimate matter in motion.

From Cosmic Organism to World Machine

In most cultures, traditional assumptions about the life of the natural world are simply taken for granted. But in ancient Greece, for the first time in Europe, these tacit assumptions were discussed and made explicit. Greek philosophers developed a sophisticated conception of nature as a living organism that was inherited by our medieval ancestors. Although there was much debate over the details, animism was central to Greek thinking. The great philosophers believed the world of nature was alive because of its ceaseless motion. Moreover, because these motions were regular and orderly, they said that the world of nature was not only alive but intelligent, a vast animal with a soul and a rational mind of its own. Every plant or animal participated psychically in the life process of the world's soul, intellectually in the activity of the world's mind, and materially in the physical organization of the world's body.[17]

In medieval Europe, Greek theories of nature, Roman technology, local, pre-Christian traditions, and the Christian religion were all brought together in an astonishing synthesis, most impressively manifested in the great gothic cathedrals. Often built at ancient sacred sites and oriented to the rising sun, they were heirs to a temple-building tradition that went back to the megalithic era. The soaring columns and vaults recall sacred groves, and vegetation bursts out everywhere. Imps, demons, dragons, and animals abound, and angels fly above. Again and again we find the mysterious figure of the Green Man, a severed head entwined with vegetation, often sprouting branches from his mouth, sometimes made of leaves (Fig. 2.1).

The orthodox philosophy of nature, taught in the cathedral schools and medieval universities, was animistic: all living creatures had souls. The soul was not in the body; rather, the body was in the soul, which permeated all parts of the body.[18] Through its formative powers, it caused the embryo to grow and develop, so that the organism assumed the form of its species. For example, an acorn sprouted into a seedling and grew into an oak tree because it was drawn or attracted toward the mature form of the tree by its soul, the soul of the oak.[19]

In animals, a further kind of soul underlay sensory perception and behavior, and gave rise to the instincts: the animal or sensitive soul. Our English word *animal* comes, obviously enough, from *anima*, Latin for "soul." In human beings, in addition to the animal instincts, there was the rational aspect of the soul: the mind or intellect. This added the qualities of thinking and free choice to those aspects of the soul that were shared with animals and plants. The human intellect was not separate from the animal and vegetative souls; rather, the rational mind was linked to animal and bodily aspects of the same soul, which were generally unconscious. In other words, the human soul included both a person's conscious mind, or spiritual essence, and the life of the body, senses, bodily activities, and animal instincts.[20]

This understanding of the human soul clearly connected human life with the life of animate nature, as well as defining the differences between plants, animals, and human beings. At the same time, man was a microcosm of the entire cosmic organism, the macrocosm (Fig. 2.2). Human society likewise reflected the hierarchical order of the universe, and the movements and conjunctions of the planets were connected with human lives and

Figure 2.1. A thirteenth-century Green Man in Bamberg Cathedral, Germany. (From Anderson and Hicks, 1985.)

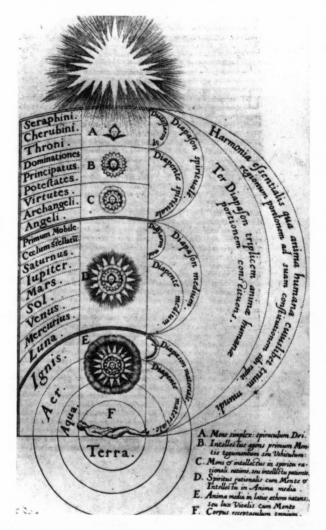

Figure 2.2. The relation of the human soul to the heavenly spheres and angelic hierarchies, according to Robert Fludd (1574–1637). This relation took place through the subtle organism whose levels (recalling the *chakras* of the Hindu system) Fludd described as follows:
(A) Pure Mind: the aperture to God; (B) the Active Intellect: the first sheath or vehicle of Mind; (C) the Rational Spirit, containing Mind and Intellect and open to Reason or Intellect; (D) the Middle Soul, containing the Rational Spirit, Mind, and Intellect; (E) the Vital Light in the Mind, or the Middle Soul swimming in ethereal fluid; (F) the Body, receptacle of all things. (From Fludd, 1619.)

the destinies of nations. Conversely, disorder in the heavens was reflected by disorder on Earth.

> . . . when the planets
> In evil mixture to disorder wander,
> What plagues and what portents, what mutiny,
> What raging of the sea, shaking of earth,
> Commotion in the winds, frights changes horrors,
> Divert and crack, rend and deracinate
> The unity and married calm of states
> Quite from their fixture.
>
> (William Shakespeare, *Troilus and Cressida*,
> Act 1, Scene 3, 93–101)

The Copernican revolution in astronomy, far from overthrowing the ancient idea of the cosmic organism, was in fact inspired by it. When Copernicus proposed that the sun, rather than the Earth, was at the center of the cosmos, he did so both because the geometrical order of the planetary spheres seemed more harmonious and because of his mystical reverence for the Sun:

> Who, in our most beautiful temple, could set this light in another or better place, than that from which it can at once illuminate the world? Not to speak of the fact that not unfittingly do some call it the light of the world, others the soul, still others the governor.[21]

Like Copernicus, Kepler regarded the center as the most fitting place for the sun, which he thought of as the prime mover of the universe. The sun "alone appears, by virtue of his dignity and power, suited for this motive duty and worthy to become the home of God himself."[22] His laws of planetary motion were embedded in a lengthy attempt to express the music of the spheres in musical notation. Like Copernicus, he was trying to express the living order of the universe in a new, more precise way; he was not challenging the idea of the cosmic organism or the idea of unseen connections between the heavens and the earth. He was in fact one of the leading astrologers of his age.[23]

The Copernican revolution began by replacing one model of the cosmic organism with another, but it soon led to the realization that the cosmos was not a closed system with a center. Rather, it was a universe with no center at all; the stars were themselves suns, and space stretched outward in all directions to infinity. The cosmic organism had broken open. Then, through the mechanistic revolution, the old model of the living cosmos was replaced by the idea of the universe as a machine. According to this new theory of the world, nature no longer had a life of her own: she was soulless, devoid of all spontaneity, freedom, and creativity. Mother Nature was no more than dead matter, moving in unfailing obedience to God-given mathematical laws.

This new worldview was first articulated on November 10, 1619, in Germany, at Neuburg on the Danube. There the twenty-three-year-old René Descartes had a visionary experience; he "was filled with enthusiasm, and discovered the foundations of a marvellous science."[24] Believing that his mystical vision was inspired by the Mother of God, he vowed to undertake a pilgrimage of gratitude to the shrine of Our Lady of Loreto in Italy, a promise he fulfilled some three years later.

The universe of Descartes was a vast mathematical system of matter in motion. Matter filled all space; it was the universal matrix. In a subtle form it swirled around in vortices; he thought the Earth and other planets were carried around the sun in such a whirlpool. Everything in the material universe worked entirely mechanically according to mathematical necessities. His intellectual ambition was boundless; he applied this new mechanical way of thinking to everything, even plants, animals, and man. Although the details of his system were soon superseded by the Newtonian universe of atomic matter moving in a void, he laid the foundations for the mechanistic worldview in both physics and biology. In the philosophy of Descartes, souls were with-drawn from the whole of the natural world; all nature was in-animate, soulless, dead rather than alive. The soul was also with-drawn from the human body, which became a mechanical automaton, leaving only the rational soul, the conscious mind, in a small region of the brain, the pineal gland. Since the time of Descartes, the favored region has moved a couple of inches to the cerebral cortex, but essentially the idea remains the same. The mind somehow interacts with the machinery of the brain, although how the two are related is still an impenetrable mystery.

In practice, the "ghost in the machine" is usually pictured as a little man inside the brain controlling the machinery of the body. Descartes did so himself. He compared the nerves to water pipes, the cavities in the brain to storage tanks, the muscles to mechanical springs, breathing to the movements of a clock. The organs of the body were like the automata in seventeenth-century water gardens, and the little man was the fountain keeper.

> External objects, which by their mere presence stimulate its sense organs . . . are like visitors who enter the grottos of these fountains and unwittingly cause the movements which take place before their eyes. For they cannot enter without stepping on certain tiles which are so arranged that if, for example, they approach a Diana who is bathing they will cause her to hide in the reeds, and if they move forward to pursue her they will cause a Neptune to advance and threaten them with his trident . . . or other such things according to the whim of the engineers who made the fountains. And finally, when a rational soul is present in this machine it will have its principal seat in the brain, and reside there like the fountain keeper who must be stationed at the tanks to which the fountain's pipes return if he wants to produce, or prevent, or change their movements in some way.[25]

The little man who operates the controls has taken a variety of forms over the years, adapting to the latest fashions in technology. A few decades ago, he was usually a telephone operator in the telephone exchange of the brain, and he saw projected images of the external world as if he were in a cinema (Fig. 2.3). Now he is often a computer programmer. Or, in an updated version of Plato's vision of the rational soul as a charioteer, he is the pilot of a jet plane. In a current exhibit at the Natural History Museum in London called "How You Control Your Actions," you look through a perspex window in the forehead of a model man. Inside is a cockpit with banks of dials and controls,

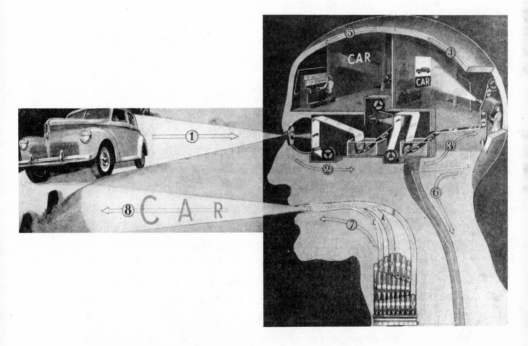

Figure 2.3. "This is what takes place in the eye, brain and larynx when we see a motor car, recognize it as such, and pronounce the word 'car.'" From a book of popular science entitled *The Secret of Life: The Human Machine and How it Works* (Kahn, 1949).

and two empty seats—presumably for you, the pilot, and your copilot in the other hemisphere. Although this Cartesian duality remains inexplicable and the "mind-body problem" is the basis of a substantial academic industry, Descartes's assumption is still generally taken for granted: The human body, like the rest of the material world, is supposed to be entirely mechanical and in principle explicable in mechanistic terms.

According to Descartes, the development and instinctive behavior of animals involved no formative agencies such as souls. Rather, animals were constructed from material particles in the seed and somehow arranged themselves by virtue of the mathematical laws of motion. In his own words:

> If we possessed a thorough knowledge of all the parts of the seed of any species of animal (e.g. man), we could from that alone, by reasons entirely mathematical and certain, deduce the whole figure and conformation of each of its members, and, conversely, if we knew several peculiarities of this conformation, we could from these deduce the nature of its seed.[26]

Descartes's visionary hope is still, more than three and a half centuries later, the dream of many mechanistic biologists. The material parts of the seed that are now supposed to determine the form of the organism are the genes. In principle, from a mechanistic point of view, "a theory of development would effectively enable one to compute the adult organism from genetic information in the egg" (L. Wolpert and J. Lewis, 1975). In fact, no mechanistic explanation of the development of even a simple plant or animal has so far been achieved, but the belief that such explanation is possible *in principle* remains a fundamental article of the mechanistic faith.[27] I return to a discussion of the mechanistic theory of life in Chapter 5.

Descartes's doctrine that plants and animals were mere machines furthered his explicit aim of making men "lords and possessors of nature."[28] Animals were automata like clocks, capable of complex behavior but lacking souls. Descartes himself dissected the heads of animals, trying to find a physical explanation of imagination and memory,[29] and through vivisection, he studied the pumping mechanism of their hearts:

> If you slice off the pointed end of the heart in a live dog, and insert a finger into one of the cavities, you will feel unmistakably that every time the heart gets shorter, it presses the finger, and every time it gets longer it stops pressing it.[30]

Since animals were supposedly inanimate, men could be free from "any suspicion of crime, however often they may eat or kill animals."[31] Hence there need be no further doubts about man's right to exploit the brute creation. Some of Descartes's followers explicitly denied that animals could feel pain; the cry of a beaten dog no more proved that it suffered than the sound of an organ proved that the instrument felt pain when struck.[32] And indeed from the time of Descartes onward, there was a great increase in the practice of vivisection.

Of course, such theories did not go unchallenged and have been widely debated ever since. One English philosopher told Descartes that his doctrine of animals as machines was "murderous"; others rejected it on the grounds that it was "against all evidence of sense and reason" and "contrary to the common-sense of mankind."[33] Within botany and zoology, the mechanistic theory of life was challenged from the seventeenth century onward by vitalists, who maintained that plants and animals were truly alive, animate in a way that machines were not.[34] There was a vitalist revival at the beginning of the present century, and it was not until the 1920s that the mechanistic theory achieved its present supremacy within academic biology.

Scientific Detachment from Nature

To this day, scientists pretend that they are rather like disembodied minds. Unlike other human activities, science is supposed to be uniquely objective. Scientific papers are conventionally written in an impersonal style, seemingly devoid of emotion. Conclusions are meant to follow from facts by a logical process of reasoning like that which might be followed by a computer if machines with sufficient artificial intelligence could ever be constructed. Nobody is ever seen *doing* anything; methods are *followed*, phenomena are *observed*, and measurements are *made*, preferably with instruments. Everything is reported in the passive voice. Even schoolchildren learn this style and practice it in their laboratory notebooks: "A test tube was taken . . ."

All research scientists know that this process is artificial; they are not disembodied minds uninfluenced by emotion. The reality is very different. To take an extreme case, here is a description of the recent race to find new kinds of superconductors:

> Thousands of researchers—my colleagues and myself included—found themselves in a scientific frenzy triggered by the prospect of fame and fortune. Day-to-day routines were marred by laboratory espionage and scientific fraud, thinly disguised threats and bald-faced lies, and an unprecedented, undignified rush to patent every minor finding, as researchers vied for Nobel Prizes and billions of dollars. (Robert Hazen, 1989)[35]

There have of course been some saintly scientists, like Michael Faraday, but most have been all too human, including Isaac Newton, who spent years in bitter disputes with Gottfried Leibniz over priority in the invention of the infinitesimal calculus.

In the mythology of science, great men are seen as archetypal heroes, endowed with superhuman powers, remembered as a glory to their nation and all humanity, their marble statues in the Hall of Fame.

> A hero ventures forth from the world of common day into a region of supernatural wonder: fabulous forces are there

encountered and a decisive victory is won: the hero comes back from this mysterious venture with the power to bestow boons on his fellow men. (Joseph Campbell, 1956)[36]

Behind the myth of the hero is the shaman, whose disembodied spirit could travel into the underworld in animal form or fly into the heavens like a bird. Like the spirit of the shaman, the mind of the scientist travels far up into the sky; he can look back from the heavens and observe the earth, the solar system, and the entire universe as if from outside.[37] He can travel in the other direction down into the most minuscule recesses of matter. In his heroic quest for truth, journeying beyond the frontiers of knowledge into the unknown, he overcomes all obstacles and returns bringing knowledge and power to mankind. This archaic image of the disembodied journey is alive and well today. "Even as he sits helplessly in his wheelchair, his mind seems to soar ever more brilliantly across the vastness of space and time to unlock the secrets of the universe," says *Time* magazine on the cover of Stephen Hawking's best-selling book *A Brief History of Time*.

The tradition of the disembodied journey is the mythic basis of scientific detachment; it is what makes it exciting, and it has a long magical and animistic tradition behind it. Imaginary journeys naturally appeal to the imagination. Descartes's philosophy grew up against this background, and his own imagination knew no bounds. But he restricted its realm to the intellect and viewed the soul as nothing but the conscious mind detached from the body and from nature itself. Ever since Descartes, a broader conception of the human soul has had to be reinvented again and again by admitting that most of its activity is unconscious.[38] In the light of Freudian and Jungian psychology, Descartes's denial of the bodily and unconscious dimensions of the psyche cannot but seem simplistic.

In his *Meditations*, he took as the first principle of his philosophy "I am thinking, therefore I exist," and his first inference was that his thinking mind was essentially disembodied:

> I saw that while I could pretend that I had no body and that there was no world and no place for me to be in, I could not for all that pretend that I did not exist. . . . From this I knew that I was a substance whose whole essence or nature is simply to think, and which does not require any place, or depend on any material thing, in order to exist. Accordingly, this "I"—that is, the soul by which I am what I am—is entirely distinct from the body, and indeed is easier to know than the body, and would not fail to be whatever it is, even if the body did not exist.[39]

His mind was therefore godlike and immortal. He could know the laws of nature through his reason and thus participate in the mathematical mind of God himself. Imagining his true self as a disembodied observer rather than an embodied participant in a living world, he provided the philosophical basis for the ideal of scientific detachment.[40] As radical feminists point out, this fantasy is characteristically male and is reinforced by the fact that most scientists are men.[41] But the ideal of scientific detachment is not confined to the ranks of professional scientists and technocrats; it has an all-pervasive influence on modern society, deepening the divisions between man and nature, mind and body, head and heart, objectivity and subjectivity, quantity and quality.

The new evolutionary cosmology has moved a long way from the world machine of classical physics. But it shares its mathematical quality; it gives us a model universe that is soundless, colorless, tasteless, odorless, and of course lifeless. This universe differs from the one we know through our senses; indeed it is inaccessible to the senses and knowable only through mathematical reason. But what kind of reality do such mathematical models represent? Do they correspond to an objective mathematical order that is more real than the world we know through our senses, as most physicists believe? Or are they simply models in our own minds that help us to understand restricted aspects of the world around us?

Sounds, smells, colors, and feelings are nowhere to be found in mathematical physics because they are excluded from the start. Physics abstracts from the world only those features that can be treated mathematically, such as shape, size, position, motion, mass, and electric charge; it ignores everything that cannot be quantified. This procedure is fundamental to physics, and it was made plain by Galileo in the early seventeenth century. Physics need take into account only the mathematical aspects of things, their "primary qualities"; these alone are regarded as objective. Other qualities known through the senses, "secondary qualities," are merely subjective, part of bodily experience; they do not exist in the objective mathematical world knowable to a disembodied mind. In Galileo's words:

> I think that these tastes, odours, colours, etc., on the side of the object in which they seem to exist, are nothing else than mere names, but hold their residence solely in the sensitive body; so that if the animal were removed, every such quality would be abolished and annihilated.[42]

The practical successes of mechanistic science bear testimony to the effectiveness of this method; the quantitative aspects of the world can indeed be abstracted and modeled mathematically. But such models leave out most of our living experience; they are a partial way of knowing. Nevertheless, the prestige that this method has acquired through physics has established it as the model of scientific detachment, the envy of biologists, sociologists, economists, and all those who aspire to scientific objectivity.

The Conquest of the Earth: The Map Becomes the Territory

Although for scientists the conquest of nature has a largely metaphorical meaning, it is only too literal for the appropriators of virgin lands, mining and logging companies, and developers in general. We see the process going on all over the world today — for example, in the virgin forests of the Amazon, Malaysia, Alaska, and the Pacific Northwest. We have exported our ideology of conquest everywhere, together with the technology that makes it possible.

The most dramatic example of this process of transformation was the opening up of the American West. It took place with a speed that amazed everyone. Into the abundant fertile lands of the West moved relentless waves of speculators and settlers. Before them retreated the wilderness and the native peoples who had lived so lightly on their sacred land. In the 1860s, as the railroads reached out westward, meat was needed, and there were buffalos by the million. They were slaughtered wholesale or gunned down for pleasure; the supplies seemed limitless. Improved rifles were invented and more deadly methods of hunting. A great hide industry sprang up, and at its height, between 1872 and 1874, over three million buffalo were hunted to supply it. By 1880, though at first no one could believe it, the buffalo were gone. For a few years more their bleached remains were a source of profit as mountains of bones were shipped to glue factories and fertilizer plants.[43] By the end of the century, fewer than a thousand of these animals survived in reservations, the pathetic remnants of the fabulous herds that only a few decades before had contained thirty to fifty million.

A similar fate befell the Plains Indians. The Plains, the last stronghold of the so-called savages, had to be cleared before settlers could feel safe and the nation achieve its destiny. After the Civil War, the guns were deliberately swiveled west under the direction of General William Tecumseh Sherman (whose middle name was, ironically, that of a great Indian prophet brutally murdered by the whites). In the early 1860s, he outlined his plan in a letter to his brother:

> The more we can kill this year, the less will have to be killed the next war, for the more I see of these Indians the

more convinced I am that all have to be killed or maintained as a species of pauper. Their attempts at civilization are simply ridiculous.[44]

By 1890, with the massacre at Wounded Knee, Sherman's dream had been accomplished.

Stripped of their myths and stories, the lands sacred to the native peoples were no longer a gift of the Great Spirit to be held in common; they became real estate. The conquered territory was divided up and bought and sold as private property. In the older settled regions, such as New England, boundaries were often related to natural features; like all traditionally settled countryside, the human divisions of the land were connected to the landscape. The territory came before the map. Not so in the virgin lands of the West.

In the rationalist spirit of the Founding Fathers, government officials superimposed a kind of Cartesian graph paper on maps, dividing them into many squares of equal size, and then into squares within squares. In due course, the map became the territory. Throughout the Midwest and the West, the square boundaries of townships, properties, and fields marched on regardless of the lie of the land, unrelated to the actual features of the place.

A new symbolic landscape was superimposed on the old. But whereas the old one was animistic, related to the spirit of the place, the new one symbolized the imposition of a rational order upon the untamed wilderness and its division into private property. Much the same happened in Canada, Australia, New Zealand, and other territories conquered and settled by Europeans. It is still going on today, as forests are divided up on maps and then destroyed in rectangles.

Indeed the same general process is typical of all development projects. First come the romantic explorers, then the scientific mapmakers, producing abstractions of the physical features of the place, detached from myth and the experience of the native peoples. Then, in air-conditioned offices, the plans are drawn up for development—road building, logging, mining, dam construction, settlement, whatever. The old animistic order, the old relationship of the native peoples to the land, is superseded as the bulldozers move in and the new order is imposed on the face of the earth.

The scientific and technological conquest of nature expresses a mentality of domination that had been widespread in the ancient world but was vastly increased in power by technology and amplified by the belief in unlimited progress. Meanwhile, the mechanistic theory of nature has taken the place of Christian missionaries in justifying the dispossession of native peoples and the disregard of their sacred places. Since nature is inanimate, their animistic relationship to the living world around them must be superstitious, their attitudes backward. They cannot be allowed to stand in the way of progress. And now, like the buffalo hunters, we can hardly believe what we have done.

THREE

RETURNING TO NATURE

The Need to Return

Returning to nature feels like going home, or reconnecting with the source of life. But few people want to return to nature for too long at a time. After all, we are the inheritors of a culture and a way of life that emphasizes our separation. We are the lords of creation, the conquerors of nature. All the ancient fears are still in the background to haunt us: the breakdown of civilization, famine, pestilence, barbarism. Our political and economic systems help to separate us from the destructive powers of nature and human nature, from forces that arouse our most basic fears, from the ever-present threat of chaos. A variety of theories and habitual attitudes reinforces and extends this primary distancing, especially the habits of scientific detachment. And then the greater the sense of separation from nature, the greater the need to return.

Nature has a variety of meanings and inspires different attempts to return. For the rationalists of the eighteenth century, she was a rational system of order, most clearly reflected in the Newtonian motions of the celestial bodies. Nature was uniform, symmetrical, and harmonious. She could be known by all mankind through reason; she was indeed the very basis of reason and esthetic judgment.

> First follow Nature, and your judgement frame
> By her just standard, which is still the same:
> Unerring Nature, still divinely bright,
> One clear, unchang'd and universal light.
> (Alexander Pope, 1711)[1]

But as the eighteenth century wore on, nature came to be understood in an almost opposite sense. She was irregular, asymmetric, inexhaustibly diverse. The change in fashion was expressed in England first through landscape gardening. Instead of clipped and manicured formal gardens, the landscaper sought to imitate an ideal of natural wildness. One of the models for the new style was found in paintings of pastoral scenes; another was in Chinese gardening:

> Writers who have given us an account of China, tell us that the inhabitants of that country laugh at the Plantations of our Europeans, which are laid out by rule and line; because they say anyone may place trees in equal Rows and uniform Figures. They choose rather to show a Genius in Works of Nature, and thereby always conceal the Art by which they direct themselves. (Joseph Addison, 1712)[2]

There was a comparable shift in attitudes to wild landscapes themselves. Previously, forests, mountains, and wild places had been seen as disagreeable and dangerous. In the seventeenth century, travelers frequently referred to mountains as "terrible," "hideous," and "rough."[3] Even at the end of the eighteenth century, most Europeans found wild, uncultivated wilderness totally unpleasing: "There are few who do not prefer the busy scenes of cultivation to the greatest of nature's rough productions," observed William Gilpin in 1791.[4] Dr. Johnson wrote of the Scottish Highlands that "an eye accustomed to flowery pastures and waving harvests is astonished and repelled by this wide extent of hopeless sterility."[5]

The new taste for wild nature was a sophisticated response, inspired to a large extent by literary and artistic models. Indeed, scenes were called landscapes because they were reminiscent of painted landscapes; they were picturesque because they looked like pictures; they were romantic because they recalled the imaginary world of romances, set far away and long ago. By the beginning of the nineteenth century, many educated people, free from the need to work on the land and encouraged by the ease of travel, attached an unprecedented importance to visiting wild and romantic places:

> Within the last thirty years, a taste for the picturesque has sprung up; and a course of summer travelling is now looked upon to be essential. . . . While one of the flocks of fashion migrates to the sea-coast, another flies off to the mountains of Wales, to the lakes in the northern provinces, or to Scotland; . . . all to study the picturesque, a new science for which a new language has been formed, and for which the English have discovered a new sense in themselves, which assuredly was not possessed by their fathers. (Robert Southey, 1807)[6]

This change of attitude was made fully explicit by the Romantic poets. For all his interest in the lives of Cumbrian shepherds, William Wordsworth believed that it took education, social station, and a long course of training to instill a taste for barren rocks and mountains. He opposed the construction of a railway to the Lake District on the grounds that it would flood the area with the urban poor, who could derive no good from immediate access to the lakes. Instead they should practice with Sunday excursions into nearby fields.[7]

By the beginning of the nineteenth century, the Romantic taste for wild nature led to an abhorrence of human interference. The attempt to improve nature destroyed it, even in landscape gardening. The painter John Constable wrote in 1822: "A gentleman's park is my aversion. It is not beauty because it is not nature."[8] Romantic nature was best experienced in solitude, and part of the attraction of the wilderness was its remoteness from the bustle of cities and industrial activity. By the mid-nineteenth century, solitude in natural surroundings was seen by many as essential for the spiritual regeneration of city dwellers. Some wilderness should be preserved both for the sake of the individual and for the sanity of society as a whole. The utilitarian philosopher John Stuart Mill, writing in 1848, already sounds modern:

> Solitude in the presence of natural beauty and grandeur is the cradle of thoughts and aspirations which are not only good for the individual, but which society could ill do without. . . . Nor is there much satisfaction in contemplating the world with nothing left to the spontaneous activity of nature; with every foot of land brought into cultivation, which is capable of growing food for human beings; every flowery waste or wild pasture ploughed up, all quadrupeds or birds which are not domesticated for man's use exterminated as his rivals for food, every hedgerow or superfluous tree rooted out, and scarcely a place left where a wild shrub or flower could grow without being eradicated as a weed in the name of improved agriculture.[9]

Clearly, the conflict between economic development and conservation was already apparent in the nineteenth century. There were even some notable conservationist triumphs, such as the preservation of Hampstead Heath in London by act of Parliament in 1871, the culmination of a long and bitter struggle against developers and their financial interests.[10] But it was in North America that wilderness took on its grandest dimensions.

American Wilderness

The early settlers in North America thought of the wilderness as practically boundless. Even in the late eighteenth century, there was no sense of its imminent conquest by man; it could be taken for granted, there was so much of it. There seemed to be room in America for everybody. "Many ages will not see the shores of our great lakes replenished with inland nations, nor the unknown bounds of North America entirely peopled," wrote a well-informed American in 1770.[11] There were limitless lands to develop—and no sense that nature was sacred or of any value in its wild state. This bounteous land needed to be improved and used by man, and only then would it be truly beautiful.

A romantic sense of nature grew up in America, as in Europe, under literary and artistic influence. One of its most influential exponents was Ralph Waldo Emerson, whose essay "Nature" (1837) transmitted a new vision of man's relationship with the world around him. The land of America expressed the same living spirit as the body of man; instead of Americans trying to impose their own historically determined consciousness on the mindless matter of the wild, they could recognize their true, living relation to the land. Emerson had a new vision, which is still visionary today: American history could be the history of the reassumption of alienated man to nature, rather than his war with it.

Like Wordsworth, Emerson recognized that this reverential attitude to nature was rare:

> To speak truly, few adult persons can see nature. . . . The
> lover of nature is he whose inward and outward senses are

truly adjusted to each other; who has retained the spirit of infancy even into the era of manhood. . . . In the woods . . . a man casts off his years as the snake his slough, and at what period soever of life is always a child. In the woods is perpetual youth. Within these plantations of God, a decorum and a sanctity reign, a perennial festival is dressed, and the guest sees not how he should tire of them in a thousand years. In the woods we return to reason and faith. There I feel that nothing can befall me in life,—no disgrace, no calamity (leaving me my eyes), which nature cannot repair. Standing on the bare ground . . . . the currents of the Universal Being circulate through me; I am part or parcel of God.[12]

By the 1850s, the opening of railways and the acceleration of economic development had made the virgin lands of America far more accessible; the wilderness could no longer be taken for granted. A disciple of Emerson's, Henry David Thoreau, was one of the first to sense the threat to virgin nature. He proposed, in vain, that each town in Massachusetts should save a 500-acre piece of woodland that would remain forever wild.

Thoreau's books on nature continually contrast the mundane and materialistic attitudes of his fellow citizens with the living world around them. Writing about the Maine woods, for example, he reflects on the logger's attitude to great trees:

The character of the logger's admiration is betrayed by his very mode of expressing it. If he told all that was in his mind, he would say, it was so big that I cut it down and then a yoke of oxen could stand on its stump. He admires the log, the carcass or corpse, more than the tree. . . . The Anglo-American can indeed cut down, and grub up all this waving forest. . . . but he cannot converse with the spirit of the tree he fells, he cannot read the poetry and mythology which retire as he advances.[13]

Thoreau was not opposed to logging, settling, and cultivation, but he thought they should be practiced in moderation and in continual relationship to wilderness preserved in the vicinity. His own experience of living in the woods near his native town of Concord, Massachusetts, was not one of complete withdrawal from society, nor did he give up living in a house. In fact, he was like the model pioneer: "I borrowed an axe and went down to the woods by Walden Pond, nearest to where I intended to build my house, and began to cut down some tall arrowy white pines, still in their youth, for timber."[14] He built a cabin, and planted a bean field in cleared land, from which he hauled out the tree stumps himself. But although Thoreau tried to live in harmony with nature, he was in continual conflict with the narrowly utilitarian attitudes of his fellow citizens. Only in an antisocial state of solitude could nature be experienced with full intensity:

> The indescribable innocence and benificence of Nature,— of sun and wind and rain, of summer and winter,—such health, such cheer, they afford forever! . . . Shall I not have intelligence with the earth? Am I not partly leaves and vegetable mould myself?[15]

Such solitary experiences of nature were heightened by the sense that the very virginity of the wilderness was threatened; the worship of nature and the urge to defend her went together.

The greatest of the Emersonian lovers and defenders of wild nature was John Muir, founder of the Sierra Club and chief protector of Yosemite Park. Muir was estranged from Christianity by his strict Presbyterian upbringing, but he was not estranged from religion. He found it in nature. The wilderness was an expression of God; man was part of nature; and nature, the fount of the world, remained his natural home. His hikes in the arduous wildernesses of the High Sierra were filled with a sense of joy. "I will touch naked God," he wrote while climbing a glacier.

And when lunching on a crust of bread: "To dine with a glacier on a sunny day is a glorious thing and makes common feast of meat and wine ridiculous. The glacier eats hills and sunbeams."[16]

In 1869—the very year that Muir began to describe the western mountains as God's outdoor temples—the transcontinental railway was completed. Muir was able to be romantic about the mountains partly because the country had been so much tamed that he could borrow or buy food within a day's walk of most of his camping places in the Sierras.[17] Now the wilderness was threatened, and Muir's response was to lobby politically for the preservation of large tracts as national parks.

The inspiration for the formation of the national parks and other reservations of wilderness was essentially religious. But in the secular and democratic spirit of American legislation, the law creating the first national park, Yellowstone, in 1872, declared it to be "set aside as a public park and pleasuring ground for the enjoyment of the people."[18] There was no mention here of sacred ground, but there was no conflict either, for pleasure and joy were part of the religion of nature. By the end of the century, such reservations of scenery were being described as "the cathedrals of the modern world."[19] They recalled, on a grander scale, the sacred groves of the ancient world, including those in Canaan.

Like temples, cathedrals, and sacred groves, the official sanctuaries of wild nature are separated from the secular world around them where other attitudes prevail. And like temples, cathedrals, and sacred groves, those who visit them can come as pilgrims—or just as tourists.

The Poet within the Scientist

In the late eighteenth and early nineteenth centuries, while the philosophy of materialism was growing in influence and conviction, the Romantics were turning to nature and finding in her life the essence of divinity. Romanticism and materialism grew up together.

For Wordsworth, nature was alive and benign. He felt in her a moral and a spiritual presence, molding and working on his mind; he was in communion with a vast invisible presence. And in his poetry, many scientists found inspiration. The name of the scientific journal *Nature* was inspired by Wordsworth, and it took as its epigraph his lines: "To the solid ground of Nature trusts the Mind that builds for aye."[20] The first issue of *Nature*, in 1869, opened with a collection of aphorisms by the German poet Goethe, expressing his vision of nature's living powers:

> Nature! We are surrounded and embraced by her; powerless to separate ourselves from her, and powerless to penetrate beyond her. . . . We live in her midst and know her not. She is incessantly speaking to us, but betrays not her secret. . . . She has always thought and always thinks; though not as a man, but as Nature. . . . She loves herself, and her innumerable eyes and affections are fixed upon herself. She has divided herself that she may be her own delight. She causes an endless succession of new capacities for enjoyment to spring up, that her insatiable sympathy may be assuaged. . . . The spectacle of Nature is always new, for she is always renewing the spectators. Life is her most exquisite invention; and death is her expert contrivance to get plenty of life.[21]

It was T. H. Huxley who chose Goethe's aphorisms to inaugurate the new journal:

> When my friend, the Editor of *Nature*, asked me to write an opening article for his first number, there came into my mind this wonderful rhapsody on "Nature" which has been a delight to me from my youth up. It seemed to me that no more fitting preface could be put before a Journal,

which aims to mirror the progress of that fashioning by Nature of a picture of herself, in the mind of man, which we call the progress of science. . . . [I]t may be, that long after the theories of the philosophers whose achievements are recorded in these pages, are obsolete, that vision of the poet will remain as a truthful and efficient symbol of the wonder and mystery of Nature.[22]

Charles Darwin too was inspired in his youth by the direct poetic experience of nature. "In connection with pleasure from poetry I may add that in 1822 a vivid delight in scenery was first awakened in my mind, during a riding tour on the borders of Wales, and which has lasted longer than any other aesthetic pleasure," he wrote in his *Autobiography*.[23] His favorite reading was Milton's *Paradise Lost*, which he took with him everywhere during his voyage on the *Beagle*.[24] But this early inspiration faded with age: "I wholly lost, to my great regret, all pleasure from poetry of any kind."[25] And with the loss of this source of inspiration, his increasingly materialist philosophy of nature was accompanied by a sense that his own thinking had become mechanical: "My mind seems to have become a kind of machine for grinding general laws out of large collections of facts," he lamented toward the end of his life.[26]

I do not know how many contemporary biologists, like T. H. Huxley, retain the poetic sense of nature experienced in their youth; there are certainly some.[27] How many are like Darwin and lose it? How many experience their minds as mechanical? How many take no delight in nature? There are no statistics. But I suspect that poetical or mystical experience of the life of nature is still a source of inspiration for many life scientists, even if half forgotten.

The Hidden Goddesses of Darwinism

Darwin turned the Romantic vision of the creative power of nature into a scientific theory. He rejected the God of the Newtonian world-machine, who was thought by Protestant theologians like William Paley to have designed and made the machinery of all living things. Instead of the heavenly Father, Darwin saw in Mother Nature the source of all forms of life. Nature herself gave rise to the Tree of Life (Fig. 3.1). Through her prodigious fertility, her powers of spontaneous variation, and her powers of selection, she could do everything that Paley thought God did. With his customary honesty, Darwin himself remarked: "For brevity's sake I sometimes speak of natural selection as an intelligent power. . . . I have, also, often personified the word Nature; for I have found it difficult to avoid this ambiguity."[28]

Darwin advised his readers to forget the implications of such turns of phrase. But if we remember instead what the personification of nature implies, we see her as the Mother from whose womb all life comes forth and to whom all life returns. She is prodigiously fertile, but she is also cruel and terrible, the devourer of her own offspring. Her fertility impressed Darwin deeply, but he made her destructive aspect the primary creative power; natural selection, working by killing, was "a power incessantly ready for action."[29]

Thus through Darwin's theory, nature took on the creative powers of the Great Mother, powers quite unsuspected in the original mechanistic conception of nature. Evolutionary philosophers conceived of these creative powers in a variety of ways. In the dialectical materialism of Marx and Engels, the creative mother principle is matter, undergoing a continual, spontaneous process of development, resolving conflicts and contradictions in successive syntheses. In the philosophy of Herbert Spenser, progressive evolution itself was the supreme principle of the entire universe. The vitalist philosopher Henri Bergson attributed the creativity of evolution to a vital impetus, the *élan vital*. In his view, the evolutionary process is not designed and planned in advance in the mind of a transcendent God but is spontaneous and creative:

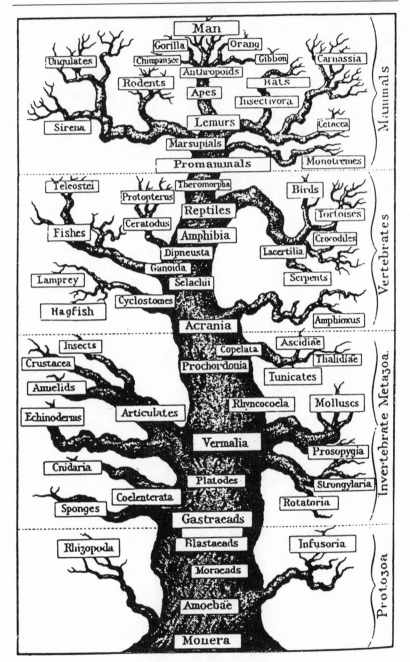

Figure 3.1. The evolutionary Tree of Life. (From Haeckel, 1910.)

Before the evolution of life . . . the portals of the future remain wide open. It is a creation that goes on for ever in virtue of an initial movement. This movement constitutes the unity of the organized world—a prolific unity, of an infinite richness, superior to any that the intellect could dream of, for the intellect is only one of its aspects or products.[30]

The neo-Darwinian theory of evolution shares this vision of evolution as a vast spontaneous creative process. As the molecular biologist Jacques Monod expressed it in his lucid summary of the neo-Darwinian worldview, *Chance and Necessity*: "Evolutionary emergence, owing to the fact that it arises from the essentially unforeseeable, is the creator of *absolute* newness." What Bergson attributed to the *élan vital*, Monod ascribed to "the inexhaustible resources of the well of chance," expressed through random mutations in DNA.[31]

In Monod's conception, the creative role of chance, the indeterminate, is expressed in its interplay with necessity, the determinate. When these abstract principles are personified, Necessity is what the poet Shelley called the "all-sufficing power" and "mother of the world." She is also Fate or Destiny, often represented by the three Fates, the stern spinning women who spin, allot, and cut the thread of life, dispensing to mortals their destiny at birth. In neo-Darwinism, the thread of life appears in a microscopic but curiously literal form in the helical DNA molecules of the genes, arranged in threadlike chromosomes.

Chance is an aspect of the goddess Fortune. The turnings of her wheel confer both prosperity and misfortune. She is the patroness of gamblers, Lady Luck, to whom we still make unconscious offerings ("one for luck . . ."). The goddess Fortune is blind. And so is chance—in Monod's words, "pure chance, absolutely free but blind, at the very root of the stupendous edifice of evolution."[32]

Maybe, as secular humanists believe, ancient conceptions of

the Great Mother and other goddesses have been superseded by modern science. But on the other hand, perhaps much of the emotional appeal of Darwinism arises from these archaic feminine archetypes, which may gain rather than lose power when they work below the surface of conscious thought.

Materialism and the Mother

Materialism in its philosophical sense asserts that only matter is real and that everything, including human consciousness, can be explained in terms of matter. As a political doctrine, it places the highest value on material well-being and material progress. In its everyday sense, it refers to a preoccupation with material needs and desires rather than spiritual values. In all these senses, the material world is the sole reality, or at least the only reality of importance.

Behind materialism in all its forms lies the figure of the Great Mother, as material reality, as Mother Nature, as the economy, as the welfare state. She is also the environment—enclosing and containing us, the source of nourishment, warmth, and protection, but we are also utterly at her mercy,[33] for the environment is uncaring and merciless; it devours and destroys.

Although many materialists have a romantic side and implicitly acknowledge the life of nature in their private lives, most of them explicitly deny it, adopting the conventional view of mankind as the only truly conscious, purposeful species in an otherwise inanimate world. From their point of view, the maternal metaphors that pervade materialist thought may tell us something about the way our minds work but have no relevance to nature itself because nature is inanimate and mechanical.

The mechanistic theory of nature has acquired such prestige through the successes of science and technology that it now seems less like a theory than a proven fact. But as science itself develops, the mechanistic worldview is being progressively transcended. Nature is coming to life again within scientific theory.

And as this process gathers momentum, it becomes increasingly difficult to justify the denial of the life of nature. For if the cosmos is more like a developing organism than a machine running down, if organisms themselves are more like organisms than machines, if nature is organic, spontaneous, creative, then why go on believing that everything is mechanical and inanimate?

One powerful reason for sticking to the mechanistic view is that it is easier; it is still the orthodoxy of industrial civilization. But it may not be easier for long. Public attitudes are greening, old political and economic certainties melting away. Doubts about the mechanistic approach to agriculture and medicine are growing; the vision of conquering nature is losing its glamour; and the climate is changing, both literally and metaphorically.

Perhaps the strongest reason for denying the life of nature is that admitting it has such overwhelming consequences. Personal intuitive experiences of nature can no longer be kept in the sealed compartment of private life, dismissed as merely subjective, for they may indeed be revelations of living nature herself, just as they seem to be at the time. Mythic, animistic, and religious ways of thinking can no longer be kept at bay. Nothing less than a revolution is at hand.

PART TWO

THE REBIRTH OF NATURE IN SCIENCE

FOUR

THE REANIMATION
OF THE
PHYSICAL WORLD

The Denial of the Life of Nature

In the scientific revolution of the seventeenth century, nature was denied the traditional attributes of life, the capacity for spontaneous movement and self-organization. She lost her autonomy. The souls that animated physical bodies in accordance with their own internal ends were exorcised from the mechanistic world of physics. Matter was inanimate and passive, acted upon by external forces in accordance with the mathematical laws of motion.

One way of understanding this crucial transition is in terms of a distinction, originally made in the Middle Ages, between *natura naturata* (natured nature), and *natura naturans* (naturing nature). The former refers to nature in the sense of what is produced, the phenomena we observe with our senses. The latter refers to the unseen productive power that gives rise to the phenomena. In the animistic physics of the Middle Ages, souls played the role of naturing nature; they organized the autonomous development and behavior of organisms, and motivated them by attraction. As we saw earlier, a seedling was attracted toward the form of the mature plant; the vegetative soul, active but invisible, gave form to the matter of the growing plant and organized it in accordance with its own ends. Stones fell to earth because they were attracted to their proper place; they were striving to go home. Souls were not outside nature according to Aristotle and his medieval followers; they were physical in the sense that they were part of nature, *phusis*.[1] When the founders of mechanistic science expelled souls from nature, leaving only passive matter in motion, they placed all active powers in God. Nature was only *natura naturata*. The invisible productive power, *natura naturans*, was divine rather than physical, supernatural rather than natural.

But this attempt to remove all traces of spontaneous organizing activity from nature ran into grave difficulties from the outset. The ghosts of the invisible souls remained, in the form of invisible forces. Gravitational attraction, acting at a distance, showed that there was more to the physical world than mere passive matter in motion. The nature of light remained mysterious, and so did chemical, electrical, and magnetic phenomena. In this chapter, I discuss how physics has progressively transcended the mechanistic theory of nature.

Gravitational Attraction

Newton's gravitational forces were inevitably mysterious. The entire universe was filled with invisible forces far more extensive than the material bodies they acted on. Through these forces, all bodies in the universe were related to all other bodies, and were somehow held in equilibrium. Everything was interconnected.

Before Newton's conception of gravitation, the interrelation of all bodies in the universe was ascribed to the soul of the universe, the *anima mundi*, or to swirling vortices of subtle substance, the "ether." Neither was material in any usual sense of the word, nor were the attractive forces with which Newton replaced them. His gravitational equation enabled the magnitude of the forces to be calculated, but it did not explain their nature.

One thing Newton firmly believed was that matter itself could not be the source of these attractive powers:

> 'Tis inconceivable that inanimate brute matter should (without the mediation of something else which is not material) operate upon and affect other matter without mutual contact. . . . That gravity should be innate inherent and essential to matter so that one body may act upon another at a distance through a vacuum without the mediation of anything else by and through which their action or force may be conveyed from one to another is to me so great an absurdity that I believe no man who has in philosophical matters any competent faculty of thinking can ever fall into it.[2]

Newton considered explanations involving subtle ethereal matter but rejected them. Such invisible matter could only interfere with the movements of the heavens that he had calculated assuming a vacuum. As far as he was concerned, the less matter the better: "The heavens are to be stripped as far as may be of all matter, lest the motions of the planets be hindered or rendered irregular."[3] In the spirit of mechanistic science, he rejected the idea of a world soul. That left only God, and he concluded that gravitational forces were a direct expression of God's will: "There exists an infinite and omnipresent spirit in which matter is moved according to mathematical laws."[4]

From the point of view of his continental critics, loyal to the Cartesian idea of vortices of subtle matter, Newton was reintroducing into nature "occult qualities," hidden causes reminiscent of souls. His use of the word *attraction*, with its animistic and sexual associations, aroused deep suspicions. Voltaire, on visiting London in 1730, thought that this was the main reason why Newton's theory was still not generally accepted in Paris: it "irritated the human mind":

> If Newton had not used the word *attraction* in his admirable philosophy, everyone in our Academy would have opened his eyes to the light; but unfortunately he used in London a word to which an idea of ridicule was attached in Paris; and on that alone he was judged adversely.[5]

As time went on, the mysterious nature of gravitational attraction was more or less forgotten. People got used to the idea, and, in spite of Newton's objection that the concept was absurd, inanimate brute matter came to be endowed with power of attraction, acting at a distance. It was not until Einstein's theory of gravitation that this mysterious attractive power received an explanation in terms of a physical but nonmaterial entity—the gravitational field.

Like the *anima mundi*, Einstein's gravitational field is not *in* space and time; rather, it *contains* the entire physical world, including space and time. The gravitational field *is* space-time, and its geometrical properties are the cause of gravitational phenomena; it acts as a formative or formal cause, like the souls of medieval philosophy. Whereas Newton's followers supposed that the attractive forces of gravitation arose inexplicably from material bodies and spread out in all directions through space, in modern physics the gravitational field is primary: it underlies both material bodies and the space between them. For example, the moon does not go round the earth because it is pulled toward it by a force, as in Newtonian physics, but because the very space-time in which it moves is curved. This model of the cosmos is nothing like the doctrines of nineteenth-century materialism, which made "inanimate brute matter" the primary reality and source of invisible forces.

Souls and Fields

The introduction of the concept of electromagnetic fields in the nineteenth century began to put back into physics spontaneously self-organizing entities with most of the traditional properties of souls. In the present century, the field concept has been extended to gravitation and to the matter fields of quantum physics, making fields more fundamental than matter.

The history of theories of magnetism illustrates the way fields have replaced souls as invisible organizing principles. The founder of ancient Greek philosophy, Thales, maintained that magnets were animate,[6] and the animistic theory of magnetism continued to predominate in Western thought well into the seventeenth century. The idea was that an invisible influence extended around the lodestone, with the power of moving matter. This invisible motivating power was a soul, as opposed to matter, and hence magnets had souls, as did electric bodies such as rubbed amber. As the soul departed from the bodies of plants and animals when they died, so magnets and electric bodies could lose their magnetic or electric powers and become inanimate again. Conversely, a magnet could induce the power of attraction in a piece of iron attracted to it, and this newly formed magnet could retain its powers for some time. The magnetic soul could therefore be transmitted, just as the life principles of plants and animals are transmitted to their offspring.

While Europeans were thinking about magnets and amber in this animistic manner, so were the Chinese, who had started using lodestones for divination at least by the beginning of the Christian era. By the eleventh century, they were using magnetic needles as compasses for navigation; this practice began in Europe in the twelfth century, and was probably transmitted from China.[7] Clearly the pointing of magnetic needles to the north meant that their magnetism was in some way related to the earth or the heavens, but how? Some thought that they were attracted to the North Pole of the heavenly sphere; others, that they were attracted to magnetic mountains near the North Pole of the earth.

In the thirteenth century, the Frenchman Pelegrinus made a spherical magnet from a lodestone and placed a compass needle at different places on its surface. He marked the lines on which the needle set itself. When the surface of the sphere had been covered with such lines, their pattern became obvious: they formed circles that girdled the magnet in the same way that meridians of longitude girdle the earth. At two points all the lines converged, just as all the meridians pass through the North and South Poles of the earth. Struck by this analogy, he called these points the poles of the magnet. He observed that the way magnets set themselves and attracted each other depended solely on the position of their poles, as if these were the seat of the magnetic power. He showed that unlike poles attracted each other and that like poles repelled. He also found that when a magnet was divided, the pieces became new magnets with new poles.[8] But he did not conclude that the earth was a magnet; he thought the attractive influence on the needle of a compass came from the Pole Star.

The founder of the modern science of magnetism, William Gilbert, took just this step. In his great book *De Magnete*, published in 1600, he proclaimed that the earth itself was a giant magnet. The dip of the compass needle showed that the magnetic influence was coming from the earth, not the heavens. The way the compass needle declined from true north, the deviation varying at different latitudes, also pointed to the earth as the source. Gilbert adopted from Pelegrinus the use of spherical magnets, and for him these "little earths" (*terrellae*) were models of the earth itself. He believed that the "true magnetic potency" of the earth was connected with its spherical form and its rotation. He adopted the traditional animistic view of magnetism. Magnetism was a sympathetic quality, and "the magnetic force is animated or is similar to soul."[9] Carefully weighing pieces of iron before and after magnetization, he found that there was no change; the soul of the magnet was weightless, just like other kinds of soul.

Gilbert followed the ancient philosophers of Egypt, Chaldea, and Greece, who believed in the existence of the soul of the universe, as well as souls of stars, planets, and the earth. He saw the powers of magnets as derived from the earth itself, describing each magnet as "an animated stone, that is a part and beloved offspring of the animated mother, Earth." His theory of magnetism was embedded in a strong sense of the life of the earth, which again and again he referred to as "the common mother" of all things.[10] Only iron and magnets are "the true and most intimate parts of the Earth," because "they retain the first faculties in nature, the faculties of attracting each other, of moving, and of adjusting by the position of the world and the terrestrial globe."[11] In his cosmological speculations, Gilbert conjectured that magnetic forces were somehow related to the earth's gravity, seeing both as aspects of the soul of the earth.

Descartes and his followers tried to get rid of such animistic conceptions by explaining magnetic and electrical phenomena mechanically in terms of flows of subtle matter called effluvia. But in the course of the eighteenth century, it became clear that there were no such material emanations (for example, the attraction between separated magnets or electric bodies was not diminished by blowing the hypothetical effluvia away). By the end of the eighteenth century, electric and magnetic effects were described mathematically in terms of an inverse square law, the effect diminishing in proportion to the square of the distance. Like gravitation, they were assumed to involve an action at a distance with nothing in between.[12] Gilbert's souls had not been explained mechanically; they had merely been replaced by inexplicable connections with no known physical basis.

Michael Faraday's researches in the first half of the nineteenth century on the relations between electrical and magnetic phenomena were part of an ambitious attempt to explain all physical reality in terms of a single kind of force pervading all space.

He wanted to explain matter in terms of converging forces, rather than forces in terms of matter.[13] In this sense, Einstein's field theories and current attempts to conceive of a primal unified field are further developments of the same grand vision that motivated Faraday, who introduced the scientific concept of the field. In so doing, he filled the theoretical void left by the mechanistic revolution of the seventeenth century with its denial of the souls of the universe; the planets, plants, and animals; and magnets and electric bodies. The subsequent history of science has involved a progressive extension of the field concept to all the natural phenomena that used to be explained in terms of souls.

Faraday thought of fields in terms of patterns of forces, as exemplified by the lines of force around a magnet (Fig. 4.1). Such fields of force were not material in any normal sense of the word, so what was their physical nature? He was not sure and considered two possible explanations. Either they were states of a subtle material medium "which we may call aether." Or they were states of "mere space." He preferred the latter, more radical explanation because of his ambition to explain all physical phenomena in terms of patterns and vibrations of forces ex-

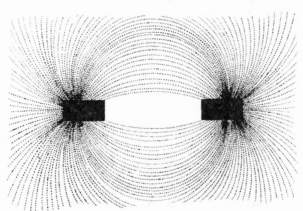

Figure 4.1. The magnetic field around the two poles of a horseshoe magnet, revealed by the pattern taken up by iron filings. (From an eighteenth-century etching.)

tended in space.[14] However, the physicist James Clerk Maxwell adopted Faraday's less favored view, and treated electromagnetic fields as states of the ether, regarded as a subtle fluid. His electromagnetic field equations unified the phenomena of electricity, magnetism, and light, and were based on a conception of lines of force as rotating tubelike vortices of ether. In Maxwell's model of the field, the ether was a mechanical medium for transmitting activity. But it soon became clear that the field had no such mechanical basis.

At the beginning of this century, the failure of experimental attempts to detect the ether as the medium of transmission of light led Einstein to explain electromagnetic phenomena in terms of fields alone. For him the ether became "superfluous." In his special theory of relativity, the electromagnetic field permeates space but has no mechanical basis whatsoever; nevertheless, it is the medium of complex processes and, like matter, has energy and momentum. It can interact with matter and exchange energy and momentum with it. But the field is independent of matter. It is not a state of matter, but of space.[15] In the general theory of relativity, Einstein then extended the field concept to gravitational phenomena. As we have already seen, the gravitational field is a space-time continuum curved in the vicinity of matter, and gravitation is a consequence of the curvature of the field.

In quantum theory, entities such as protons and electrons are regarded as wave packets, or quanta of vibration. They exist as vibrations of quantum matter fields, one kind of field for each kind of particle: a proton is a quantized pattern of vibration in the proton-antiproton field; an electron a quantum of the electron-positron field, and so on. These matter fields are states of space, or of the vacuum. But the vacuum is not empty; it is full of energy and undergoes spontaneous fluctuations that can create new quanta "from nothing." A particle and its antiparticle can spring into "virtual existence" at a point in space and then immediately annihilate each other. "A vacuum is not inert and featureless, but alive with throbbing energy and vitality" (Paul Davies, 1984).[16] The atoms of old-style materialism—hard, permanent particles of matter moving in a void—are nowhere to be found. Atoms, like other quantum systems, are structures of activity rather than changeless inert things.

The result of all these changes is that fields, together with energy, have become the basis of physical reality. In the phrase of Karl Popper, through modern physics, "materialism has transcended itself."[17]

Universal Energy

In the animate world of medieval philosophy, there were two fundamental ingredients in nature: matter, which in itself was indeterminate and chaotic, and souls, which gave form to material bodies and organized and motivated all physical activity. Matter, in this Aristotelian sense, was quite unlike the atomic particles of Newtonian physics. It was a universal principle whose inherent nature was indeterminate; it was pure potentiality and could take on any form. The modern conception of a single universal energy, capable of existing in many different forms, has given physics a unitary principle that has more in common with the Aristotelian conception of matter than anything in Newtonian physics.

The seventeenth-century mechanical universe of matter in motion possessed a variety of ingredients that were separate from each other: matter itself, made up of indestructible, passive atoms; motion; the attractive forces of gravitation; electrical and magnetic forces; light; the forces of chemical combination; and heat. The modern conception of energy provides a unifying principle for them all.

The first great step toward this synthesis was taken around the middle of the nineteenth century, with the introduction of the general concept of *energy* (a term not previously used in physics), together with the principle of energy conservation. One ingredient was the seventeenth-century concept of *vis viva* ("living force"), defined as the mass times the square of the velocity of a moving body, mv^2. (This corresponds closely to the modern concept of kinetic energy, $mv^2/2$.) Another came from the study of electrical and magnetic forces, particularly Faraday's contributions toward a unified conception of force. Yet another came from the study of heat engines by the founders of thermodynamics, who originally thought in terms of a subtle fluid called caloric that flowed from hot bodies to cold. The formulation of a unified concept of energy enabled all these kinds of force, flow, activity, and potentiality to be related to one another. At the same time, the connection of energy with mechanical work was made clear, and the distinction between actual and potential energy was established. Maxwell built this new conception of energy into his electromagnetic theory and thus brought light and other forms of electromagnetic radiation into the new synthesis.[18]

By the 1890s, some physicists, such as Wilhelm Ostwald, were claiming that not matter but energy was the sole real substance in nature.[19] These claims were viewed with suspicion by most physicists (after all, matter itself was still thought of in terms of permanent atoms). But the final step was taken by Einstein, who through his famous equation $E = mc^2$ (where E is energy, m mass, and c the velocity of light), established the equivalence of mass and energy, and the fact that they can be converted to each other.

The result is that all nature is now thought to consist of fields and energy. Energy, like Aristotelian matter, can exist in many different forms. In Aristotelian physics, these forms were organized by souls; in modern physics, they are organized by fields.

Indeterminism and Chaos

For three centuries, from the time of Descartes until 1927, physicists lived under the spell of a powerful illusion. Everything was believed to be fully determinate and hence in principle, though not in practice, entirely predictable. In the early nineteenth century, this illusion was epitomized by the French physicist Pierre Laplace in his fantasy of a demon able to calculate all the past and future:

> Consider an intelligence which, at any instant, could have a knowledge of all forces controlling nature together with the momentary conditions of all the entities of which nature consists. If this intelligence were powerful enough to submit all these data to analysis it would be able to embrace in a single formula the movements of the largest bodies in the universe and those of the lightest atoms; for it nothing would be uncertain; the future and the past would be equally present before its eyes.[20]

Laplace's demon was not an omniscient God but a superhuman scientist. He was a disembodied intelligence capable of godlike knowledge through mathematical calculation, an idealized physicist with a mind that could embrace all nature—in fact, an idealized Laplace.[21]

In 1927, as quantum theory developed, it became clear that at the microscopic level, physical processes were essentially indeterminate, predictable only in terms of probabilities. For several more decades, such randomness was assumed to have little relevance to the everyday world. But over the last twenty years or so, it has become generally recognized that indeterminism is inherent in systems at all levels of complexity: in "dissipative processes" far from thermodynamic equilibrium, where small fluctuations can be amplified to produce large effects (as in the formation of convection cells in a heated fluid);[22] in "catastrophic" processes such as the breaking of waves, the turbulent flow of liquids, and phase transitions (as in boiling or freezing); in the weather; in living organisms; in brains; in population dynamics and ecology; and in the behavior of the economy. These kinds of processes cannot be modeled effectively in terms of old-style deterministic physics. New mathematical approaches are required, the most important of which is "chaos theory."[23] In mathematical models of chaotic processes, made possible by computers, the model systems do not settle down into a simple equilibrium but develop in complex, nonrepetitive patterns (Fig. 4.2).

In retrospect, it is clear that even in the abstract world of Newtonian mechanics, with its point masses, perfectly elastic parti-

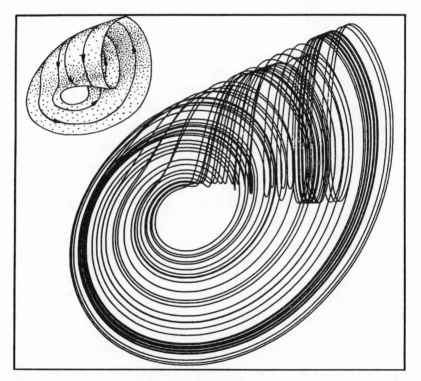

Figure 4.2. A chaotic attractor, the Rössler funnel, discovered by Otto Rössler, one of the pioneers of chaotic dynamics. Chaotic attractors, like other kinds of attractors, exist in "phase space," in which the state of a dynamical system at a given instant is represented by a point. As the system changes, the point moves, tracing out a trajectory. In chaotic attractors these trajectories do not settle down into simple repetitive patterns, but continue to behave chaotically. (From Abraham and Shaw, 1984.)

cles, frictionless wheels, and other mathematical fictions, most physical systems follow no exactly predictable paths.[24] Even very simple systems consisting of coupled pendulums behave chaotically.[25] So does the solar system itself, long regarded as the centerpiece of deterministic physics.

> Unfortunately, non-chaotic systems are very nearly as scarce as hen's teeth, despite the fact that our physical understanding of nature is largely based upon their study. ... Algorithmic complexity theory and nonlinear dynamics together establish the fact that determinism reigns only over a quite finite domain; outside this small haven of order lies a largely uncharted, vast wasteland of chaos. (Joseph Ford, 1983.)[26]

Thus an inherent spontaneity in the life of nature has once again been recognized by science, after a denial lasting over three hundred years. The future is not fully determined in advance; it is open. Insofar as it can be modeled mathematically, it has to be modeled in terms of chaotic dynamics. And this chaos, openness, spontaneity, and freedom of nature provide the matrix for evolutionary creativity.

Attractors

One of the principal purposes of the mechanistic theory of nature was to expel all internal purposes from the natural world. In the animistic physics of the Middle Ages and the Renaissance, organisms developed toward their own ends; their activity was purposive in that it was attracted toward their internal goals. Such teleological explanations were supposedly banished from the mechanical world of Newtonian physics, although this world was still pervaded by mysterious forces of attraction. The course of nature was not pulled toward future ends or goals, but only pushed from the past. But in recent decades, the ends toward which physical systems develop have been reinvented in the form of dynamic *attractors*.

Dynamics is the branch of science concerned with the forces that change or produce the motion of bodies. With the development of the modern geometrical theory of dynamical systems, the concept of the attractor has become central. The mathematical models contain structures called *basins of attraction* within each of which is a nucleus, the attractor. The dynamic system—which can be a pendulum, for example, or a developing embryo—evolves toward the attractor, which defines the end toward which it will habitually move.

For example, think of a pudding basin as a basin of attraction. The bottom of the basin is the attractor. Small balls can be thrown into the basin from any angle and at any speed. They will move around in different patterns, depending on the way they are thrown in. But all of them will end up at the attractor, the bottom of the basin. Dynamic models are like abstract versions of such basins, existing in mathematical spaces called *vectorfields*. "The vectorfield is a model for the habitual tendency of the situation to evolve from one state to another, and is called the dynamic of the model."[27]

Some attractors are points, like the bottom of a pudding basin; the dynamical system comes to rest in a particular end-state. Others are cycles: the system settles down into a repetitive periodic behavior (for example, the pendulum of a clock). And some attractors, called *strange attractors*, are chaotic (Fig. 4.2); the system never settles down into an exact repetitive pattern.

Although mathematical models of attractors are not conventionally regarded as teleological, they inevitably imply the existence of ends or goals, even if these ends are "chaotic." In such models, the vectorfield plays the formative role of the soul, and the attractor is the end toward which any dynamic system within the basin of attraction is habitually attracted. Not surprisingly, such models have been applied to the development of embryos, animal behavior, and other biological phenomena.[28] Here, for the first time since the mechanistic revolution, is a kind of mathematical modeling that can represent the purposive features of living organisms, interpreting their behavior in terms of ends, not just mechanical forces pushing them from the past. (More on this subject in the next chapter.)

The Mystery of Dark Matter

One of the most surprising and puzzling features of the universe revealed by modern cosmology is that most of the matter in it is utterly unknown to us—"dark matter." The existence of dark matter was first inferred from a study of the behavior of galaxies and the halos of gas around them. If galaxies contained only the stars and the gas we can observe, plus a generous allowance for other known forms of matter, their gravitational attraction would not even begin to explain their observed behavior, such as the way they cluster together. The way that galaxies and their halos hang together can be explained only if there is a great deal of hidden matter within and between them. This dark matter has powerful gravitational effects, but its constitution is unknown.

Recent estimates of the amount of dark matter in the universe vary from around 90 to 99 percent. In other words, the familiar kinds of matter we know about make up only 1 to 10 percent of the total, less than the tip of an iceberg. Some of this dark matter may be dark remnants of stars, including black holes; most of it probably consists of exotic kinds of particles, unlike any actually detected by nuclear physicists. Since these dark particles scarcely interact with familiar material things, with our senses or our instruments, they are by their very nature elusive. They are all around us without our knowing it. Their properties are still a matter of speculation, and physicists are currently trying to find ways of detecting them, for example through very subtle magnetic interactions they might occasionally take part in.[29]

The magnitude of this mystery is staggering. The great majority of the matter in the universe is utterly unknown, except through its gravitational effects. Yet through the gravitational field, it has shaped the way in which the universe has developed. It is as if physics has discovered the unconscious. Just as the conscious mind floats, as it were, on the surface of the sea of unconscious mental processes, so the known physical world floats on a cosmic ocean of dark matter.

This dark matter has the archetypal power of the dark, destructive Mother. It is like Kali, whose very name means "black." If there is more than a critical amount of dark matter, then the cosmic expansion will gradually come to an end, and the universe begin to contract again, pulled in by gravitation, until everything is ultimately devoured in a terminal implosion, the opposite of the Big Bang—the Big Crunch.

The Rebirth of Nature in Science

Nature is now once again seen to be self-organizing. Instead of the soul of the universe and all the other kinds of soul within it, the basis of this self-organization is now seen as the universal field of gravitation and all the other kinds of fields within it. Indeterminism, spontaneity, and creativity have reemerged throughout the natural world. Immanent purposes or ends are now modeled in terms of attractors. And beneath everything, like a cosmic underworld, is the inscrutable realm of dark matter.

These developments have brought back many of the features of animate nature denied in the mechanistic revolution; in effect, they have begun to reanimate nature.[30] But of course they have not returned us to the premechanistic worldview; they are leading toward a postmechanistic worldview, at a higher turn of the spiral. For the modern conception of nature gives an even stronger sense of her spontaneous life and creativity than the stable, repetitive world of Greek, medieval, and Renaissance philosophy. All nature is evolutionary. The cosmos is like a great developing organism, and evolutionary creativity is inherent in nature herself.

FIVE

THE
NATURE
OF LIFE

The Life Force

The starting point for speculation about the nature of biological life is death. What happens when a plant, an animal, or a person dies? The body remains. It still weighs the same. It still has the same shape and the same material constituents. Yet it is now dead. It can no longer grow, or move, or maintain itself. It starts to decay. Something seems to have left it—the life-force, the breath, the spirit, the soul, the subtle body, the vital factor, or the organizing principle.

All over the world, people have come to similar conclusions. *Something* leaves the body when it dies. And whatever it is, it is not made of ordinary matter; it is immaterial, made of subtle matter, or it is a kind of flow, like the flow of the breath, or like fire. Flow theories of life are traditionally associated with the idea that the flow of life, the life-spirit, is not only within living organisms but all around them. The breath of life is also the air, the wind, the spirit. It is the animating principle of all nature. Here, for example, is an account of the beliefs of a tribe of Amazonian Indians:

> The Ufaina believe in a vital force called *fufaka* which is essentially masculine and which is present in all living beings. This vital force, whose source is the sun, is constantly recycled among plants, animals, men and the Earth itself which is seen as feminine. Each group of beings, men, plants, animals, Earth or water require a minimum amount of this vital force in order to live. When a being is born, the vital force enters it and the group to which the being belongs. The group is seen as borrowing the energy from the total stock of energy. When a being dies it releases this energy, and returns it to the stock. It is once again recycled. Similarly, when a living thing consumes another, for example when a deer eats a bush, a man eats a deer. . . . The consumer acquires the energy of the consumed, and it accumulates in its own body.[1]

In terms of modern science, this vital force is energy. Energy is indeed present in all things. Living organisms draw it from their environment, as plants take it from the sun in photosynthesis and animals take chemical energy from their food through digestion and respiration. They accumulate it in their own bodies and use it to power their movements and behavior. When they die, the energy accumulated in their bodies is released to continue on its way in other forms. The flow of energy on which your body and your brain depend at this very moment is part of the cosmic flux, and the energy within you will flow on after you are dead and gone, taking endless new forms.

But the vital force, spirit, or energy flow can be only one of the aspects of life. The very fact that the same energy can take so many different forms means that something else must account for these forms themselves. If the same energy can exist in the body of a plant, in the deer that eats it, and in the man that eats the deer, then the plant, the deer, and the man must owe their different characteristics to some formative principle over and above the flow of energy—indeed to a principle that organizes this flow in accordance with its own ends. Aristotle called this principle the *psyche* (the soul). He also called it *entelechy* (from *en*, meaning "in," and *telos*, meaning "end"), that which has its end in itself, its own internal purposes.

Thus life involves both an energy flow, which can be understood as an aspect of the universal energy flux, and a formative principle, which gives an organism the ends toward which its life processes are attracted. The big question concerns the nature of this formative principle. For over three hundred years, the science of life has been the arena of a long and often bitter debate on this very question.

Three Theories of Life and Nature

One tradition of thought, *vitalism*, has maintained that living organisms are truly alive, animate rather than inanimate. They are organized by immaterial souls, vital factors, formative impulses, or entelechies.[2] In essence, vitalism is a development of the animistic theory of nature that held sway in Europe before the mechanistic revolution. But, whereas the old animistic theories treated all nature as alive to some degree, vitalism has confined life to biological organisms, leaving the rest of nature to mechanistic physics.

By contrast, the mechanistic theory of life denies that there is any essential difference between living organisms and dead organisms, or inanimate matter in general. It regards organisms as inanimate machines, governed only by the general laws of nature that apply in the realms of physics and chemistry (Chapter 2). Although when they die their organization obviously breaks down, there is no difference in kind between a living and a dead organism; life does not involve any additional principles unknown to physics, only a difference of degree. Living and dead bodies obey the same universal laws of physics and chemistry. The organization of living organisms does not depend on any nonmaterial principle other than these laws; it somehow emerges from complex physicochemical interactions in a manner that remains obscure.

Ever since Descartes, the nagging problem for mechanists has been the purposiveness of living organisms. Embryos seem to have an inner urge to grow up into adult organisms, and even if they are damaged, they often manage to mature normally. And the instincts of animals—such as the building of webs by spiders or the migrations of swallows—reveal that they are motivated by inner drives and purposes. Vitalists ascribe these self-motivating characteristics of organisms to their souls, or life-principles. Mechanists deny the existence of any such entities and therefore have to find a substitute for them; the soul has to be reinvented in a mechanistic guise. In the late nineteenth century, this inner organizing principle was identified with the germ plasm inside the cell nuclei. The nucleus was like a tiny brain directing and controlling the body of the cell around it. This role is now attributed to the genes, consisting of molecules of DNA. But far from being mere inanimate molecules, the genes have been endowed with all the properties of life and mind. They are even supposed to be selfish. The living world is thought of as a capitalist economy, and the individualistic, selfish, and competitive characteristics of man taken for granted by free-enterprise economic theories are then projected onto the genes.

In the vivid language of Richard Dawkins, organisms are "throwaway survival machines" built by the selfish genes for themselves to live in. These genes are no longer mere chemicals; they have come to life and have minds like ruthless men. Not only do they have powers to "create form," "mold matter," and "choose," but they engage in "evolutionary arms races" and even "aspire to immortality."[3] The selfish-gene theory takes anthropomorphism to an extreme unprecedented in science.

The most popular contemporary machine metaphors for organisms are provided by computers and their programs. The purposive organizing principles of organisms are now commonly regarded as "genetic programs." This is another way of endowing the DNA molecules with properties of life and mind; they are like molecular souls. In the case of computer programs, the designs and purposes originate in human minds, but who writes the genetic programs? Although most biologists still claim to be mechanists, in effect the paradigm of modern biology has become a cryptic form of vitalism, in which "genetic programs" or "selfish genes" play the role of the vital organizing factors.

Whereas the mechanistic and vitalist theories both date back to the seventeenth century, a third theory, the *holistic* or *organismic* or *systems* theory, has grown up since the 1920s. This holistic approach attempts to transcend the long-standing vitalist-mechanist controversy. It agrees with the mechanists in affirming the unity of nature, seeing the life of organisms as different in degree from the rest of the physical world but not different in kind. And it agrees with the vitalists in stressing that organisms are organic wholes and cannot be reduced to the physics and chemistry of simpler systems.

The holistic theory in effect treats all nature as alive, and in this respect represents an updated version of premechanistic animism. From this point of view, even crystals, molecules, and atoms are organisms (Fig. 5.1). They are not made up of inert atoms of matter, as in old-style atomism. Rather, as modern physics itself has shown, they are structures of activity, patterns of energetic activity within fields. In the words of the philosopher Alfred North Whitehead: "Biology is the study of the larger organisms, whereas physics is the study of the smaller organisms."[4] And in the light of modern cosmology, physics is also

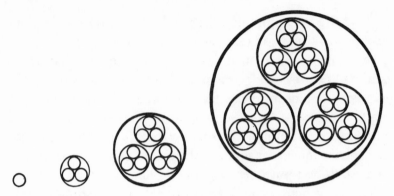

Figure 5.1. Successive levels in a nested hierarchy of organisms or holons. At each level the holons are wholes containing parts, which are themselves wholes containing lower-level parts, and so on. The diagram could, for example, represent subatomic particles in atoms, in molecules, in crystals; or cells in tissues, in organs, in organisms; or planets in solar systems, in galaxies, in galactic clusters; or phonemes in words, in phrases, in sentences.

the study of the all-embracing cosmic organism and of the galactic, stellar, and planetary organisms that have evolved within it.

The Mystery of Development and Regeneration

Mechanistic biology has been most successful in accounting for the physiology of adult organisms. It regards them as machines; their organs are like parts of a machine that function together harmoniously to maintain the organized integrity of the whole. The interrelation of the parts depends on processes of feedback: the activities of the machine are regulated by internal controlling devices that are themselves influenced by the machine's activity—for example, steam engines regulated by "governors," or thermostats, or the complex cybernetic, computerized systems that guide missiles toward their targets. The purposive functioning of the organs in relation to the organism as a whole has been shaped by nature in the course of evolution by natural selection. But it is no more due to an inherent soul or vital factor than is the activity of a guided missile. The purposes of organs and of organisms are programmed in their genes; they arise from the chemical molecules of the genetic material, DNA.

The machine analogy has some plausibility in relation to adult organisms: machines, especially those containing feedback control systems, are indeed like artificial organisms or organs: airplanes are like birds, cameras like eyes, pumps like hearts, computers like brains. Machines, made by human beings to serve human purposes, reflect some of the organic, purposeful qualities of the people that make and use them. But the fact that machines are like artificial organisms does not mean that organisms are nothing but machines.

The machine analogy breaks down when it comes to understanding the growth and development of organisms, their morphogenesis (from the Greek words *morphe*, "form," and *genesis*, "coming into being"). Oak trees grow from tiny embryos in acorns; elephants develop from small fertilized eggs that look very much like the eggs of any other mammal. No machines grow and develop spontaneously from machine eggs; they have to be assembled from preexisting parts in factories. Nor do they reproduce by giving rise to new machines from small parts of themselves; nor do they regenerate after damage. By contrast, if a flatworm is cut into pieces, each can develop into an entire flatworm (Fig. 5.2). If hundreds of cuttings are made from a willow tree, each can grow into a new tree. Pieces of wound tissue from plant stems can regenerate roots and shoots;[5] and even single cells from such wound tissue, grown in test tubes, can form entire plants.[6]

Vitalists have always argued that morphogenesis and regeneration cannot be explained mechanistically. As physical systems, machines are nothing more than the sum of their parts and the interactions between the parts. If parts are taken away, the integrity of the machine is lost. By contrast, living organisms have a wholeness that is more than the sum of the parts and their interactions. They can often regain their normal forms even if parts are removed. There is something within them that is holistic and

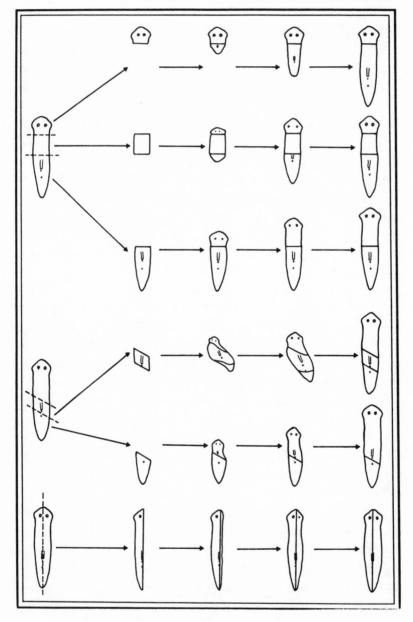

Figure 5.2. The regeneration of complete flatworms from pieces of worm, cut as indicated by the dotted lines. (After Morgan, 1901.)

purposive, directing their development toward the normal adult form of their species—the soul, or the life-principle.

A sophisticated vitalist theory was developed at the beginning of the twentieth century by the German embryologist Hans Driesch.[7] Following Aristotle, he called the nonmaterial purposive vital factor "entelechy." He thought that the entelechy of an organism somehow contained within itself the form or plan of the adult organism. It attracted the developing organism toward this end. And within the organism was a nested hierarchy of entelechies: the entelechy of the eye, for example, and within it, entelechies for its parts such as the retina and the lens. Driesch argued that the genes give rise to the chemicals that make up the organism, but the way these chemicals are ordered in cells, tissues, and organs—for example in eyes, leaves, feathers, or brains—depends on entelechies.

Driesch thought that entelechy acted by imposing order on otherwise probabilistic, indeterminate physical and chemical processes within the organism. But since he was writing at a time when physical determinism was still taken for granted, he had to suppose that entelechy itself introduced indeterminism into the physical processes within the body. This seemed like a fatal flaw in his theory, for it was unthinkable that some mysterious vital factor could interfere with deterministic physics. Ironically, by the 1920s, when the mechanistic theory had become predominant in academic biology and vitalism was treated as a discredited heresy, the development of quantum theory made Driesch's theory far more plausible. According to quantum theory, physical and chemical processes both within organisms and outside them were probabilistic anyway.

The mechanistic school of biology has always rejected vitalist arguments as a matter of principle. Entelechies and other "vital factors" are regarded as superstitious survivals from an animistic past that have no place in rational scientific discourse. The only valid scientific explanations are mechanistic explanations. But as we have seen, the trouble is that the purposive and holistic nature of morphogenesis and regeneration continues to defy mechanistic explanation, so vital factors are reinvented in mechanistic guises, such as selfish genes and genetic programs. Such programs are inherited, purposive, holistic organizing principles; they do everything that entelechies were supposed to do (Fig. 5.3). They do not consist of matter per se, but of information. And information is what puts form into things, it in-forms; it plays the same role as entelechy, but it sounds more scientific.

The computer analogy that lends the idea of the genetic program its plausibility is inherently dualistic. The programs, the software, organize the operations of the material components of the computer, the hardware. The programs are purposive; they are designed with particular ends in mind. Insofar as they have mindlike properties, they have them because they are the products of human minds. The idea of the genetic program inevitably suggests that it is mindlike and purposive, and that it acts upon the matter of the organism to organize it in accordance with its in-built ends. In this sense it provides a new metaphor for the

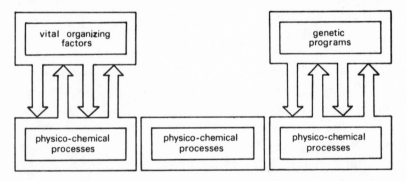

Figure 5.3. Diagrammatic representations of the vitalist (left), mechanistic (center) and modern crypto-vitalist (right) theories of life. According to vitalists, the physicochemical processes of life are organized by vital organizing factors, such as entelechy. The mechanistic theory denies the existence of such factors and asserts that life can be understood in terms of physico-chemical processes alone. But in the nominally mechanistic modern conception, these processes are organized by genetic "programs," "information," or "instructions," which play much the same role as the organizing factors of vitalism.

soul, or entelechy. But it takes us no nearer a genuine mechanistic explanation.

The genetic program as a vital factor is not the same as the DNA molecules in the genes, for these are just molecules, not mindlike entities. The fact that qualities of mind are commonly projected onto the genes, especially the qualities of selfish, competitive people within capitalist societies, makes it easy to forget that they are just chemicals. As such, they play a chemical role, and their activity is confined to the chemical level. The genetic code in the DNA molecules determines the sequence of amino-acid building blocks in protein molecules, the so-called primary structure of the proteins. The genes dictate the primary structure of proteins, not the specific shape of a duck's foot or a lamb's kidney or an orchid. The way the proteins are arranged in cells, the ways cells are arranged in tissues, and tissues in organs, and organs in organisms are not programmed in the genetic code, which can only program protein molecules. Given the right genes and hence the right proteins, and the right systems by which protein synthesis is controlled, the organism is somehow supposed to assemble itself automatically. This is rather like delivering the right materials to a building site at the right times and expecting a house to grow spontaneously.

Consider your arms and legs. They contain exactly the same kinds of muscle cells, nerve cells, and so on. They contain the same proteins and other chemicals; the bones are made of identical substance. Yet they have different shapes, just as houses of different design can be made from the same building materials. The chemicals alone do not determine the form. Nor does the DNA. The DNA is the same in all the cells of the arms and the legs, and indeed everywhere else in the body. All the cells are genetically programmed identically. Yet somehow they behave differently and form tissues and organs of different structures. Clearly some formative influence other than DNA must be shaping the developing arms and legs. All developmental biologists acknowledge this fact. But at this stage their mechanistic explanations peter out into vague statements about "complex spatio-temporal patterns of physico-chemical interaction not yet fully understood." Obviously this is not a solution but just another way of stating the problem.

In the 1960s and 1970s, a number of influential molecular biologists, encouraged by their success in "cracking the genetic code," entered the field of developmental biology with high hopes of solving the basic problems within a decade or two. The key was supposed to be an understanding of how the synthesis of proteins was controlled. But disillusionment has already set in. At a conference in 1984, Sydney Brenner summarized current thinking among developmental biologists as follows:

> At the beginning it was said that the answer to the understanding of development was going to come from a knowledge of the molecular mechanisms of gene control. I doubt whether anyone believes that anymore. The molecular mechanisms look boringly simple, and they don't tell us what we want to know. We have to try to discover the principles of organization.[8]

Brenner suggested that instead of the misleading concept of the genetic program, the organization would be better understood in terms of "internal representations" or "internal descriptions."[9] But these are surely yet other ways of conceiving of vital organizing factors.

Morphogenetic Fields

In the 1920s, in a holistic spirit, a number of biologists independently proposed a new way of thinking about biological morphogenesis: the concept of embryonic, developmental, or morphogenetic *fields*. These fields were like the known fields of physics in that they were invisible regions of influence with inherently holistic properties, but they were a new kind of field unknown to physics. They existed within and around organisms, and contained within themselves a nested hierarchy of fields within fields—organ fields, tissue fields, cell fields. As in the science of magnetism and electricity souls were replaced by electromagnetic fields, by a comparable step in biology, entelechies were replaced by biological fields.

Magnetic fields were in fact one of the principal analogies used by the proponents of morphogenetic fields. Just as cutting magnets into pieces gave rise to small but complete magnets, each with its own magnetic field, so cutting up organisms like flatworms gave rise to pieces with complete flatworm fields, enabling them to regenerate into flatworms.

Like entelechies, morphogenetic fields attract developing systems to the ends, goals, or representations contained within themselves. Mathematically, morphogenetic fields can be modeled in terms of attractors within basins of attraction.[10] The mathematician René Thom has expressed this idea as follows:

> All creation or destruction of forms, or morphogenesis, can be described by the disappearance of the attractors representing the initial forms, and their replacement by capture by the attractors representing the final forms.[11]

The idea of morphogenetic fields has been widely adopted in developmental biology. But the nature of these fields has remained obscure. Some biologists think of them as useful turns of phrase but in reality consisting of no more than "complex spatio-temporal patterns of physico-chemical interaction not yet fully understood." Others think of these fields as governed by morphogenetic field equations that exist in a Platonic realm of eternal mathematical forms. Thus the morphogenetic field equations for the dinosaurs, for example, always existed, even before the Big Bang. The equations were not affected by the evolution of the dinosaurs or by their extinction. The morphogenetic field equations for all past, present, and future species, and indeed for all possible species (many of which may never actually exist), somehow dwell eternally in a transcendent mathematical realm. These mathematical truths are beyond time; they cannot evolve and are not affected by anything that actually happens in the physical world. They are like ideal designs for all possible organisms in the mind of a mathematical God.

There is a third way of conceiving of these fields. According to the hypothesis of formative causation, they are a new kind of field so far unknown to physics with an intrinsically evolutionary nature. The fields of a given species, such as the giraffe, have evolved; they are inherited by present giraffes from previous giraffes. They contain a kind of collective memory on which each member of the species draws and to which it in turn contributes. The formative activity of the fields is not determined by timeless mathematical laws—although the fields can to some extent be *modeled* mathematically—but by the actual forms taken up by previous members of the species. The more often a pattern of development is repeated, the more probable it is that it will be followed again. The fields are the means by which the habits of the species are built up, maintained, and inherited.

Morphic Resonance

The hypothesis of formative causation, first proposed in my book *A New Science of Life* (1981) and further developed in *The Presence of the Past* (1988), suggests that self-organizing systems at all levels of complexity—including molecules, crystals, cells, tissues, organisms, and societies of organisms—are organized by "morphic fields." Morphogenetic fields are just one type of morphic field, those concerned with the development and maintenance of the bodies of organisms. Morphogenetic fields also organize the morphogenesis of molecules; for example, shaping the way the strings of amino acids coded for by the genes fold up into the complex three-dimensional structures of proteins. Likewise, the development of crystals is shaped by morphogenetic fields with an inherent memory of previous crystals of the same kind. From this point of view, substances such as penicillin crystallize the way they do not because they are governed by timeless mathematical laws but because they have crystallized that way before; they are following habits established through repetition.

The way past hemoglobin molecules, penicillin crystals, or giraffes influence the morphic fields of present ones depends on a process called *morphic resonance*, the influence of like upon like through space and time. Morphic resonance does not fall off . with distance. It does not involve a transfer of energy, but of information. In effect, this hypothesis enables the regularities of nature to be understood as governed by habits inherited by morphic resonance, rather than by eternal, nonmaterial, and nonenergetic laws.

This hypothesis is inevitably controversial, but it is testable by experiment, and there is already considerable circumstantial evidence in its favor. For example, when a newly synthesized organic chemical is crystallized for the first time (say a new drug), there will be no morphic resonance from previous crystals of this type. A new morphic field has to come into existence; of the many energetically possible ways the substance could crystallize, one actually happens. The next time the substance is crystallized anywhere in the world, morphic resonance from the first crystals will make this same pattern of crystallization more probable, and so on. A cumulative memory will build up as the pattern becomes more and more habitual. As a consequence, the crystals should tend to form more readily all over the world.

Such a tendency is in fact well known; new compounds are generally difficult to crystallize, sometimes taking weeks or even months to form from supersaturated solutions. As time goes on, they tend to appear more readily all over the world. The most popular explanation among chemists for this phenomenon is that fragments of previous crystals are transferred from laboratory to laboratory on the beards or clothing of migrant chemists.[12] They then serve as nuclei for new crystals of the same type. Or these crystal seeds are supposed to be blown all over the world as microscopic dust particles in the atmosphere. The hypothesis of formative causation predicts that such crystallizations should still occur more readily under standardized conditions as time goes on, even when migrant chemists are rigorously excluded from the laboratory and dust particles filtered from the atmosphere.

In the realm of biological morphogenesis, the hypothesis predicts that when organisms follow an unusual path of development—for example, when abnormal adults arise as a result of exposing embryos to an unusual environment—the more often this happens, the more likely it is to happen again. There is already evidence from experiments on fruit flies that they are indeed more likely to develop abnormally after others have done so.[13]

From this point of view, living organisms inherit not only genes but also morphic fields. The genes are passed on materially from their ancestors and enable them to make particular kinds of protein molecules; the morphic fields are inherited nonmaterially, by morphic resonance, not just from direct ancestors but also from other members of the species. The developing organism tunes in to the morphic fields of its species and thus draws upon a pooled or collective memory.

Genetic mutations can affect this tuning process and the ability of the organism to develop under the influence of the fields, just as changes in the condensers or other components of a TV set can affect its tuning to particular channels or affect the reception of programs; the sounds or pictures may be distorted. But just because mutant components can affect the pictures and sounds produced by the TV receiver, this does not prove that the TV programs are programmed by the set's components and generated inside the set. No more does the fact that genetic changes can affect the form and behavior of organisms prove that their form and behavior are programmed in the genes.

The Mystery of Instinct

Instinctive behavior shows the same holistic, purposive characteristics as morphogenesis. A female mud wasp, for example,
builds an underground nest, lines it with mud, and then builds
a large tube and funnel over the entrance. The function of the
structure seems to be the exclusion of parasitic wasps, which
cannot get a grip on the smooth interior of the funnel (Fig. 5.4).
She lays an egg at the end of the nest hole, which she then stocks
with paralyzed caterpillars sealed into separate compartments.
Finally, she plugs the hole at ground level with mud, destroys the
carefully constructed funnel, and scatters the fragments. She
does all this instinctively, without having to learn from other
wasps.

This sequence of behavior, like instinctive behavior in general,
consists of a series of "fixed action patterns."[14] The end point of
one serves as the starting point for the next. And as in morphogenesis, the same end points can be reached by different routes
if the normal pathway is disturbed. For example, if an almost
complete funnel is damaged, the mud wasp rebuilds it; it is regenerated (Fig. 5.4).

From a vitalist point of view, such purposive instinctive
behavior depends on the organizing activity of the soul, or entelechy, which organizes the activity of the senses, the nervous system, and the motor organs toward the achievement of its ends.
From the crypto-vitalist point of view now conventional, this
goal-directed organization can be ascribed to the genetic program. But how the synthesis of particular proteins results in
complex goal-directed behavior such as that of the mud wasp
remains utterly obscure.

From a holistic point of view, such purposive behavior depends on holistic organizing principles. The nature of the principles, sometimes called "emergent systems properties," is
usually left obscure. I think of them as morphic fields, which,
like other kinds of morphic fields, are inherited by morphic res-

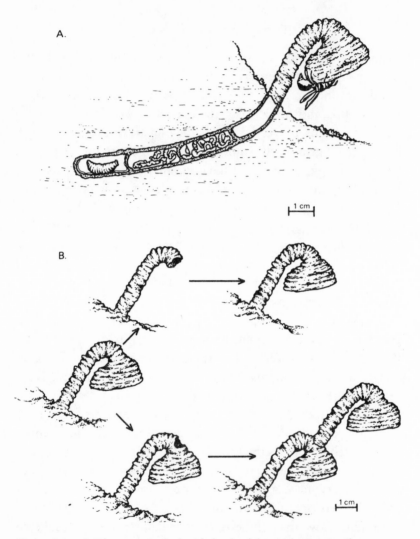

Figure 5.4. A. The nest, stocked with food, of the mud wasp *Paralastor*. B. The damaged funnels are repaired by the wasps. *Above:* a new funnel is constructed after the old one has been removed. *Below:* an extra stem and funnel are made in response to a hole above the normal funnel. (After Barnett, 1981.)

onance. Instincts are the behavioral habits of the species and depend on a collective unconscious memory. Through the morphic fields, patterns of behavior are drawn toward ends or goals provided by their attractors.

If behavior is indeed governed by morphic fields, when some members of a species acquire a new pattern of behavior and hence a new behavioral field, for example, by learning a new trick, then others should tend to learn the same thing more quickly, even in the absence of any known means of connection or communication. The more members of the species that learn it, the greater should this effect become all over the world. Thus, for example, if laboratory rats learn a new trick in America, rats in laboratories elsewhere should show a tendency to learn it faster. There is experimental evidence that this effect actually occurs.[15]

The Mystery of Memory

Even very simple animals have the capacity to learn from experience. And even the fixed action patterns of instinctive behavior often involve individual learning: mud wasps, for example, learn to recognize various features of the environment around the nest they are building; otherwise, they would be unable to find their way back to the nest when they went to fetch mud or hunt caterpillars. And learning implies memory. How do they remember?

Mechanistic theories of memory inevitably suppose that memory depends on material "memory traces" somehow stored within the nervous system. These hypothetical traces are often likened to connections in a telephone exchange, or tape recordings, videotapes, or computer memory stores. The most popular idea is that memory traces somehow depend on modifications of the junctions between nerve cells, the synapses.

Neuroscientists have been trying for decades to locate memory traces in the brains of experimental animals. The usual procedure is to train the animals to do something and then to cut out parts of their brains to find out where the memories are stored. But even after large chunks of their brains have been removed—in some experiments over 60 percent—the hapless animals can often remember what they were trained to do before the operation.[16] One researcher summarized the recurrent failure to find localized memory traces by remarking that "memory seems to be both everywhere and nowhere in particular."[17]

Some scientists have proposed that memories may be stored in a distributed manner, vaguely analogous to the storage of information in holograms, over broad regions of the brain.[18] Others postulate the existence of unidentified "backup" storage systems to account for the survival of learned habits after various putative memory stores have been removed by surgery. But there may be a ridiculously simple reason for these recurrent failures to find memory traces in brains: They may not exist. A search inside your TV set for traces of the programs you watched last week would be doomed to failure for the same reason: The set tunes in to TV transmissions but does not store them.

The hypothesis of formative causation suggests that memory depends on morphic resonance rather than material memory stores. Morphic resonance depends on similarity. It involves an effect of like on like. The more similar an organism is to an organism in the past, the more specific and effective the morphic resonance. In general, any given organism is most like *itself* in the past and hence subject to highly specific morphic resonance from its own past. For instance, you are more like you were a year ago than like I was. This self-resonance helps to maintain an organism's form, in spite of the continuous turnover of its material constituents. Likewise, in the realm of behavior, it tunes in an organism specifically to its own past patterns of activity. Neither your habits of behavior, speech, and thought nor your memories of particular facts and past events need be stored as material traces in your brain.

But what about the fact that memories can be lost as a result of brain damage? Some types of damage in specific areas of the brain can result in specific kinds of impairment: for example, the loss of the ability to recognize faces after damage to the secondary visual cortex of the right hemisphere. A sufferer may fail to recognize the faces even of his wife and children, even though he can still recognize them by their voices and in other ways.[19] Does this not prove that the relevant memories were stored inside the damaged tissues? By no means. Think again of the TV analogy. Damage to some parts of the circuitry can lead to loss or distortion of picture; damage to other parts can make the set lose the ability to produce sound; damage to the tuning circuit can lead to loss of the ability to receive one or more channels. But this does not prove that the pictures, sounds, and entire programs are stored inside the damaged components.

This idea enables the working of individual memory and the inheritance of instincts and behavioral capacities to be seen as different aspects of the same phenomenon. Both depend on morphic resonance, but the former is more specific than the latter. Individual memory and capacities for learning take place against the background of a collective memory inherited by morphic resonance from previous members of the species. In the human realm, such a concept already exists in Jung's theory of the collective unconscious as an inherited collective memory.[20] The hypothesis of morphic resonance enables the collective unconscious to be seen not just as a human phenomenon but as an aspect of a far more general process by which habits are inherited throughout nature.

The Mystery of Social Organization

Societies of termites, ants, wasps, and bees can contain thousands or even millions of individual insects. They can build large elaborate nests, exhibit a complex division of labor, and reproduce themselves. Such societies have often been compared to organisms at a higher level of organization, or superorganisms.

Inevitably, there has been a long-standing debate as to whether such societies really represent a higher level of living organization with irreducible holistic properties of their own or whether they should be regarded as aggregates explicable in terms of their parts and the mechanistic interactions between the individual insects. From a vitalist point of view, the colony as a whole has a soul that coordinates the individual insects within it.[21] By contrast, mechanists have to try to understand it all in terms of the behavior of parts studied in isolation. As a matter of principle, no mysterious soul or holistic organizing factor can be involved.[22]

From the holistic perspective, such colonies are indeed organisms at a higher level than the individual insects within them. Their organizing principles are usually conceived of in vague terms such as *systems properties* or *self-organizing patterns of information*. I propose that they should be considered as morphic fields. Such fields embrace and include the individuals within them, as magnetic fields embrace and include the particles of iron that they organize into characteristic patterns. The individual insects are within the social morphic field, just as the iron particles are within the magnetic field. From this point of view, trying to understand the social morphic field on the basis of the behavior of isolated insects would be just as impossible as trying to understand the magnetic field by taking iron filings out of it and studying their mechanical properties in isolation.

The organization of insect colonies involves several mysterious features quite apart from the prodigious complexity of the social organization itself. For example, in his studies of South African termites, the naturalist Eugene Marais found that they could speedily repair damage to the mounds, rebuilding tunnels and arches, working from both sides of the breach he had made, and meeting up perfectly in the middle, even though the individual insects are blind. He then carried out a simple but fascinating experiment. He took a large steel plate several feet wider and higher than the termitary and drove it right through the center of the breach so that it divided the mound, and indeed the entire termitary, into two separate parts:

> The builders on one side of the breach know nothing of those on the other side. In spite of this the termites build a similar arch or tower on both sides of the plate. When eventually you withdraw the plate, the two halves match perfectly after the dividing cut has been repaired. We cannot escape the ultimate conclusion that somewhere there exists a preconceived plan which the termites merely execute.[23]

From the present point of view, such a plan would exist within the morphic field of the colony as a whole. By morphic resonance, this would contain a collective memory of all similar termite colonies in the past, as well as a memory of the colony's own past, by self-resonance.

The behavior of shoals of fish and flocks of birds likewise shows a coordination that has so far defied explanation. Vast flocks of dunlins, for example, can wheel and bank as if they are a single superorganism, and the rate at which the "maneuver waves" pass through the flock is too rapid to admit of any simple mechanistic explanation. The idea of their coordination taking place through the morphic field of the flock, extending around and embracing the individual birds, seems to make better sense.[24]

In a similar way, social morphic fields can be thought of as coordinating the behavior of herds of reindeer, pods of whales, and all patterns of social organization. The same principles should apply to human societies.[25] The members of a traditional tribe, for instance, are included within the social field of the tribe and the fields of its cultural patterns. These fields have a life of their own and give the tribe its habitual patterns of organization, maintained by self-resonance with the tribe itself in the past. Thus the field of the tribe includes not just its living members but also its past members. And indeed, all over the world, the invisible presence of the ancestors in the life of traditional social groups is explicitly recognized.

The Controversies Continue

Vitalist theories of the nature of life attributed its purposive organization to nonmaterial souls, or vital factors under a variety of other names. Mechanistic theories have always denied the existence of such "mystical" entities but have then had to reinvent them in new guises. Vitalists have always criticized the reductionism of the mechanistic approach and have drawn attention to its limitations and inadequacies. Mechanists have always criticized vitalism on the grounds that it is sterile, relying on mysterious hypothetical entities unamenable to experimental investigation. By contrast, they point out, the mechanistic approach has been very productive and has led to an understanding of many aspects of organisms, such as the genetic code for proteins, which were previously completely unknown and unsuspected.

Meanwhile, for over sixty years, organicists have tried to transcend the vitalist-mechanist controversy by stressing the holistic properties of living organisms. They see biological organisms as examples of the holistic systems found at all levels of complexity, from atoms to galaxies.[26] Some organicists, especially advocates of the systems approach, have retained the machine metaphor but have adopted more sophisticated versions of it.[27] Systems theorists, partly from fear of being branded as vitalists, have generally avoided proposing that there are new kinds of causal entities in nature, like souls, or fields unknown to physics. Rather the problems are to be understood in abstract terms, such as information transfer and feedback, without bothering too much about the physical base of these processes, which are implicitly presumed to depend only on the known fields and forces of physics.[28]

Other organicists have concentrated on the idea of organizing fields, such as morphogenetic fields. Such fields to some extent demystify the old idea of souls, but at the same time they mystify the idea of fields, endowing them with surprising properties undreamed of in nineteenth-century physics. The trouble is that the nature of these fields has remained obscure. Mechanists habitually criticize them on the same grounds as vital factors: They are not amenable to experimental investigation. And this criticism is valid if morphogenetic fields are regarded as no more than a way of talking about complex but conventional physicochemical interactions or reflections of eternal mathematical truths in a transcendent Platonic realm.

However, if morphogenetic fields (like other kinds of morphic fields) are regarded as habitual, they *do* become experimentally testable. Such fields contain an inherent memory given by morphic resonance, and as such they differ from the current conception of the known fields of physics, which are still thought to be governed by eternal laws. According to the hypothesis of formative causation, morphic fields are not only at work in living organisms but in crystals, molecules, and other physical systems. These too are organized by fields with an inherent memory. Now that all nature is thought to be evolutionary, it is no longer possible to take for granted the conventional idea that all chemical and physical systems are governed by eternal laws of nature. The so-called laws of nature may be more like habits, maintained by morphic resonance.

SIX

COSMIC EVOLUTION AND THE HABITS OF NATURE

Eternity and Evolution

The recent revolution in cosmology has thrown science into crisis. It has brought our two most fundamental models of reality into collision. One of these, the paradigm of eternity, tells us that nothing really changes. The other, the paradigm of evolution, tells us that everything changes. Until recently, these theories were kept safely apart. Evolution was kept down to earth, and the heavens were eternal.

Physicists used to think that they were studying an eternal universe, governed by eternal mathematical laws, consisting of an eternal amount of matter and energy, conserved forever according to the principles of conservation of matter and energy. These fundamental realities of physics were not altered by anything that actually happened in the universe. The evolution of life on earth, for example, made no difference to them, nor would the extinction of all life on our planet. The laws of nature and the eternal quantities of matter and energy always were and always will be the same.

Meanwhile, in biology, the human sciences, politics, economics, and technology, the paradigm of progressive evolution reigns supreme. Everything changes and develops in time. This sense of development pervades our conceptions of ourselves, life as a whole, and our entire planet.

In the context of the eternal world-machine, evolution on earth was just a local fluctuation in an essentially repetitive universe going on forever in the same way; or even worse, a world-machine that was slowly running out of steam, heading toward a thermodynamic "heat death" when entropy or disorder would reach a maximum—in mythic terms, the dissolution of the cosmos into chaos. This cheerless prospect has been accepted by many twentieth-century intellectuals as an unavoidable truth, unquestionably established by science. It has provided the scientific background for a torrent of books, plays, poems, and paintings about "existential anguish," the "loss of meaning," and the "ultimate futility of life." The optimistic idea of progressive development on earth was offset by a cosmic pessimism. Everything would inevitably come to an end in an exhausted universe with nowhere to go.

Like the legendary Procrustes who placed his victims on an iron bed, cutting them down to size if they were longer than the bed and if shorter stretching them, Charles Darwin tried to fit the evolution of life onto the Procrustean bed of the deterministic, mechanistic universe of nineteenth-century physics. His twentieth-century followers have gone on trying to force the evolution of life into an eternal universe. The "modern synthesis" that laid the foundations of the neo-Darwinian theory took place in the 1930s and 1940s.[1] It was an attempt to create a thoroughly mechanistic theory of biological evolution consistent with the physical sciences. All evolutionary phenomena were to be explained, in principle, in terms of the eternal laws of physics and chemistry. Making random genetic mutations the ultimate source of all evolutionary novelty allowed nature a kind of creativity consistent with a blind, purposeless, nonevolving universe.

But physics itself has since adopted an evolutionary cosmology. All nature is now seen to be evolving, not just the realm of life on earth. And this throws many of the old certainties into question. If all nature is evolutionary, then what about the laws of nature? Were eternal laws imposed upon nature from the outset, like a kind of cosmic Napoleonic Code? Or have the laws of nature evolved along with nature herself, like a universal common law? Or are the regularities of nature more like habits that have grown within the developing universe?

Cosmic Evolution

The newborn universe was filled with energy hurtling outward. As the cosmos grew and cooled, first subatomic particles, then atoms, galaxies, stars, molecules, crystals, planets, and biological life developed within it. We live in a world that was born some fifteen billion years ago, a world that has always been growing and is still growing today. On this planet, life has been developing for over three billion years in an evolutionary process that continues in ourselves. The development of science is part of this very process.

This is the modern creation story. The Big Bang is like the primal orgasm, the generative moment. Or it is like the breaking open of the cosmic egg.[2] The cosmos is like a growing organism, forming new structures within itself as it develops. Part of the intuitive appeal of this story is that it tells us that everything is related. Everything has come from a common source: all galaxies, stars, and planets; all atoms, molecules, and crystals; all microbes, plants, and animals; all people on this planet. We ourselves are related more or less closely to everyone else, to all living organisms, and ultimately to everything that is or that ever has been. One of the great themes of traditional creation myths is the division of the primal unity into many parts, the emergence of the many from the one. The modern theory of cosmic evolution fulfills this mythic role.

Another of the intuitive attractions of the modern story is its affirmation of creativity in the universe, in life, and in humanity. The creative process not only occurred long ago in the mythic time of origins; it has been going on ever since and is still going on today. This vision reinforces our modern fascination with innovation, change, and development; we can experience human creativity as part of this cosmic creative process.

The modern creation story undoubtedly reflects our cultural preoccupations, but it is difficult to know to what extent it involves projecting them onto the world around us. In the Judeo-Christian myth of history, the end (the new creation) mirrors the beginning (the first creation). For decades, we feared that humanity and much of life on this planet would be destroyed by an apocalyptic nuclear war. It was fitting that we should have a model of cosmic history in which the beginning in a vast explosion mirrored our dread of our end in a nuclear holocaust.

Moreover, in the 1960s, the idea that the universe would go on expanding forever was in gratifying harmony with the idea that the world economy would keep growing indefinitely. Now we have more doubts, and sure enough, unknown amounts of dark matter now loom beneath the surface of the visible universe threatening to end the cosmic expansion.

Laws or Habits of Nature?

Modern attempts to create a mathematical Theory of Everything are still based on a number of assumptions carried over from old-style mechanistic physics. The most important of these is that the laws of nature are eternal; they were there "before" the beginning and governed the universe from the outset. Heinz Pagels has expressed this idea as follows:

The nothingness "before" the creation of the universe is the most complete void that we can imagine—no space or time or matter existed. It is a world without place, without duration or eternity, without number—it is what mathematicians call "the empty set." Yet this unthinkable void converts itself into a plenum of existence—a necessary consequence of physical laws. Where are these laws written into that void? What "tells" the void that it is pregnant with a possible universe? It would seem that even the void is subject to law, a logic that exists prior to time and space.[3]

This way of thinking bears a strong resemblance to the traditional Christian theology of creation by the word, or *logos*, of God. The mother principle is the primal chaos, or the pregnant void. The theological notion of God-given eternal laws of nature was built into the foundations of mechanistic science, and it persists as the implicit metaphysical basis for modern cosmology. If the mind of God is dissolved away, we are left with free-floating mathematical laws playing the same role as laws in the mind of God. Stephen Hawking, for example, takes for granted this assumption of eternal laws and believes that if physicists understood "the basic laws of the creation and subsequent evolution of the universe," theoretical physics would reach its end, an end he thinks is already in sight.[4]

However, in order to entertain this ambitious idea, Hawking has to make a number of other gigantic assumptions, still shared by most physicists. One is the old reductionist doctrine that everything can ultimately be explained in terms of the physics of the smallest particles of matter:

Since the structure of molecules and their reactions with each other underlie all chemistry and biology, quantum mechanics enables us in principle to predict nearly everything we see around us, within the limits set by the uncertainty principle. (In practice, however, the calculations required for systems containing more than a few electrons are so complicated that we cannot do them.)[5]

As we have seen, this idea is of little help in understanding the fundamental problems of biology, chemistry, the weather, or the world of everyday experience (Chapters 4 and 5). It has the same illusory quality as Laplace's notion that the entire course of the universe was predictable in principle (though not in practice) on the basis of Newtonian mechanics. It is a modern version of the fantasy of mathematical omniscience.

Modern mathematical cosmology is in fact a strange theoretical hybrid between the paradigms of eternity and evolution. It retains the Pythagorean or Platonic assumption so beloved of mathematicians—the notion that everything is governed by an eternal realm of mathematical order, transcending space and time.[6] Yet it opens up a great evolutionary vision of all nature and in so doing throws its own foundations into question. If all nature evolves, why should the laws of nature not evolve as well? Why should we go on assuming that they are eternally fixed?

The idea of the laws of nature is based on a political metaphor. Just as human societies are governed by laws, so the whole of nature is supposedly ruled by the laws of nature. In the seventeenth century, the metaphor was quite explicit. God, the lord of the world-machine, had framed the laws that govern everything, and these laws existed eternally in his mathematical mind.

When we return to the source of this analogy, we see at once that human laws are not eternal; they change and develop. The present laws of the United States, for example, are not the same as they were a hundred years ago, nor will they be the same in a hundred years' time. They evolve along with the political, social, and economic systems they govern and within which they are framed. If we want to retain the metaphor of natural law, then in an evolving universe, it would make sense to think of the laws of nature evolving along with nature herself.

The trouble with the legal metaphor is that it implies an autocratic lawgiver or some kind of cosmic legislature to make the new laws. It also requires some sort of universal enforcement agency to make sure that they are obeyed. In the original version of mechanistic physics, God fulfilled these roles, and his power was so overwhelming that no particle of matter could ever disobey him. Nature had no life, power, creativity, or spontaneity of her own; her compliance with God's laws was total. But if the laws of nature are made up as evolution goes on, how are they framed and enforced?

God could still provide an answer, but he would now have to be an evolutionary God, making and enforcing new laws as the universe developed rather than thinking them eternally. When the first crystal was created, or the first protein molecule, or the first living cell, or the first bird, or the first thinking mind, he would have framed the relevant laws and then ensured that these laws applied everywhere in the universe thereafter.

Or, if we want to do without God yet persist with the idea of universal laws, we could say that the creation of the first crystal, protein, and so on involved the spontaneous appearance of the relevant laws and rules, and that these then spread everywhere instantaneously and thereafter applied universally. But it would be impossible to distinguish this view by experiment from the conventional assumption that all the laws of nature are eternal. Before a new phenomenon that is not in practice predictable on the basis of known laws, we have no way of proving whether the laws governing it existed before.

By contrast, the notion that the regularities of nature are more like habits is scientifically testable. As discussed in Chapter 5, new kinds of molecules, crystals, organisms, patterns of behavior, and patterns of thought should tend to occur more readily the more they have happened before, according to the hypothesis of formative causation. And there is already evidence that this habit-forming process actually occurs.

The habits of most kinds of physical, chemical, and biological systems have been established for millions, even billions of years. Hence most of the systems that physicists, chemists, and biologists study are running in such deep grooves of habit that they are effectively changeless. The systems behave *as if* they were governed by eternal laws because the habits are so well established. The idea that they are governed by eternal laws is an idealization or abstraction that approximates to the facts, but it is not a metaphysical truth.

In summary, there are three possible models of the regularities of nature in the context of evolutionary cosmology. First, there is the traditional model that all the laws of nature are eternal and in some sense prior to the physical universe in space and time. Second, there is the idea that new laws come into being as nature evolves and thereafter apply universally. And third, there is the idea that the regularities of nature are essentially habitual and that a kind of memory is inherent in nature. This habit model implies that past patterns of activity influence those in the present. According to the hypothesis of formative causation, this influence takes place by morphic resonance. As the habit model is distinguishable by experiment from models based on immutable laws, it should be possible to find out if it is a better model. At present, the question is open.

If memory within nature sounds mysterious, we should bear in mind that mathematical laws transcending nature are more rather than less so; they are metaphysical rather than physical. The way mathematical laws can exist independently of the evolving universe and at the same time act upon it remains a profound mystery. For those who accept God, this mystery is an aspect of God's relation to the realm of nature; for those who deny God, the mystery is even more obscure: A quasi-mental realm of mathematical laws somehow exists independently of nature, yet not in God, and governs the evolving physical world without itself being physical.

Many scientists avoid thinking about this problem by acknowledging, if challenged, that the mathematical models of physics exist only in our minds. Stephen Hawking, for example, assures us that "a theory is just a model of the universe, or a restricted part of it, and a set of rules which relate quantities in the model to observations that we make. It exists only in our minds and does not have any other reality (whatever that might mean)."[7] But if these models are only in our minds, how are we to account for the regularities of nature themselves, the repeatable phenomena that scientists study and make models of? In the face of this problem, most scientists (usually implicitly, often unconsciously) fall back on the idea that mathematical laws of nature do indeed exist independently of human minds, that they are objective realities, whether we can describe them or not. Then the problem of what these immaterial laws actually are, and how they work, is usually avoided by flipping back to the idea that they are just models in our minds.

If the laws of nature are indeed mathematical models in human minds, they may be models of habitual aspects of evolutionary nature rather than models of eternal laws. After all, in an evolutionary universe, the regularities of nature evolve: that is what evolution is all about.

Evolutionary Physics

Contemporary cosmologists and theoretical physicists are intensely preoccupied with the first few fractions of a second of the newborn universe, when all its energy became manifest, together with fundamental particles and the fundamental fields of nature.

The most popular current theory of the origins of the fields of nature, the *superstring theory*, proposes the existence of a primal unified field in ten dimensions—nine of space and one of time. As the universe expanded and cooled, the symmetries of this primal field were broken, and one by one, the known fields of physics separated from the unified field (which nevertheless continues to exist, even though its unified nature is no longer manifest). First, at around 10^{-44} seconds from the beginning, the gravitational field separated out; then, at around 10^{-36} seconds, the quantum matter fields that give rise to strong nuclear forces; then, at around 10^{-10} seconds, the electromagnetic field and the fields of the weak nuclear forces separated from each other.[8]

As we have already seen in Chapter 4, the fields of modern physics play many of the same roles as souls in animistic, pre-mechanistic philosophies of nature. In this context, it is significant that the contemporary conception of a primal unified field, a cosmic field of fields, bears a strong resemblance to the neo-Platonic conception of the world soul. The philosopher Plotinus in the third century A.D. thought of this cosmic soul as the source of all the souls within it: "There is both Soul and many souls. From the one Soul proceed a multiplicity of different souls."[9] Modern unified field theories can be paraphrased in a parallel manner: "There is both the one Field and many fields. From the one Field proceed a multiplicity of different fields." But the modern theory is of course far more detailed than old theories of the soul of the universe, and it is far more evolutionary. It describes a process of cosmic becoming, even ascribing approximate times to the various stages in which energy, fields, matter, galaxies, stars, and planets came into being as the universe grew and developed.

Why Is the Universe As It Is?

One of the profound questions raised by any cosmology is *why* the world should be organized as it is, and not otherwise. Traditionally, God provided the answer. One way of thinking of God's creative beneficence was to suppose that he could have created many other kinds of world, but owing to his goodness, he chose to create the best of all possible worlds. This argument was elaborated in the seventeenth century by the philosopher Gottfried Leibniz and is best known through Voltaire's satire of it in his novel *Candide*, where Dr. Pangloss holds firm to the doctrine that all is for the best in the best of all possible worlds, even in the face of the most ludicrous misfortunes.

In the context of modern evolutionary cosmology, this question is discussed not in terms of God but of man. One of the central facts that cosmology has to take into account is that cosmologists themselves exist. Like other human beings, they could not have existed in most of the possible universes of which physicists can conceive by altering various aspects of their equations and inserting different values for the "constants" of nature. "For example, if the relative strengths of the nuclear and electromagnetic forces were to be slightly different, then carbon atoms could not exist in nature and human physicists would not have evolved."[10] Obviously, the properties of the universe *must* be consistent with our own evolution and present existence. This fact is expressed in the form of the *cosmological anthropic principle*.

The so-called weak form of the anthropic principle is not controversial:

> The observed values of all physical and cosmological quantities are not equally probable but they take on values restricted by the requirement that there exist sites where carbon-based life can evolve and by the requirement that the Universe be old enough for it to have already done so.[11]

From here it is a short step to the *strong anthropic principle*: "The Universe must have those properties which allow life to develop within it at some stage in its history."[12] This is controversial because it implies a governing purpose in the origin and evolution of the universe, and for centuries any consideration of purpose has been outlawed by mechanistic science. One interpretation of the strong anthropic principle is that "there exists one possible universe 'designed' with the goal of generating and sustaining observers."

From the strong anthropic principle a further step leads to the *final anthropic principle*, which takes the assumption of purpose even further. Supposing that the strong anthropic principle is correct and that intelligent life must come into being at some stage in the universe's history, then if it dies out without having any significant effect on the universe as a whole, it is hard to see why it *must* have come into existence in the first place. Following this line of reasoning, the final anthropic principle asserts: "Intelligent information-processing must come into existence in the universe, and, once it comes into existence, it will never die out."[13]

Whether or not we choose to admit that the evolution of the cosmic organism is purposive, the very idea that other universes are possible raises the question of not only why this particular universe has the quantitative features that it does but how its particular features are maintained. The eternal realm of mathematics imagined by Platonists presumably contains the mathematical laws of all possible universes, so how was this subset of mathematical possibilities brought into relation to the newborn universe in the first place and maintained thereafter?

Again, God could provide one kind of answer: He designed this universe, skillfully selecting the values of the numerical constants of nature, and he then maintained them by remembering them. Alternatively, the "constants" could be remembered within nature herself rather than by a mind transcending nature. Once particular patterns were established—however they came into being in the first place—they could become increasingly habitual through repetition. Perhaps the numerical constants of physics and the properties of the known physical fields are in fact long-established habits. They could have been different, but only a universe that developed these particular habits could hang together as ours does and allow the evolution of habits of chemical, biological, cultural, and mental organization within it.

Natural Selection of the Habits of Nature

If nature is organized by eternal transcendent mathematical laws, they have to be framed with the greatest precision. The exact values of all numerical "constants" have to be specified precisely from the outset. The designing mind of a mathematical God is still there in the background. By contrast, if nature is organized habitually, the regularities of nature can grow up within the developing cosmos by an organic evolutionary process. Not all new patterns of organization that come into being in the physical, chemical, biological, cultural, and mental realms are viable. Only those that are in harmony with their environment can survive, and only through survival and repetition can they become habitual.

The evolution of habits—such as the way protein molecules fold up, crystals form, or plants develop; or the instincts of animals; or human cultural and mental habits—involves a two-stage process. First, the new pattern has to come into being by a creative leap or synthesis; second, it is subject to natural selection.

You are probably familiar with this two-stage process from your own experience. New ideas, for example, or new ways of doing things generally arise suddenly by a creative jump or insight. Then they are subject to a process of selection. Some are so successful that they become habitual; others are rejected, die out, or fade away. The same is true of biological evolution. New bodily forms can arise suddenly, for example, as a result of genetic mutations or unusual environmental conditions, and so can new patterns of behavior. These are then subject to natural selection, and the successful oft-repeated patterns become increasingly habitual. According to the hypothesis of formative causation, this happens not only because of genetic inheritance but also because of morphic resonance from previous similar organisms.

Creative jumps also occur in the chemical realm. Many new

kinds of molecules and crystals are still coming into being for the first time through the activities of synthetic chemists; they are new material forms, new patterns, new syntheses. Indeed, all existing molecules and crystals, such as the benzene molecule or the mica crystal, must likewise have come into being for the first time at some stage in the past; even atoms did not always exist. Their present forms and properties may just be successful habits. Natural selection may be operating in the atomic, molecular, and crystalline realms, just as it is in the biological realm. And if molecules and crystals do indeed inherit a memory from previous ones of their kind by morphic resonance, it should be possible to study the building-up of habits by means of experiments with newly synthesized chemicals and crystals.

Galaxies and stars also represent repetitive patterns of organization, falling into distinct types with characteristic life cycles. Perhaps these too are habitual; successful patterns of galactic and stellar organization have, through repetition, become increasingly probable. The same may be true of planetary systems and planets. Perhaps there are other planets elsewhere of the same species as Venus, say, or Jupiter, or earth. This raises the mind-boggling possibility that our own planet may be in morphic resonance with similar planets elsewhere in the universe. The evolutionary process on earth may have been following a habitual pattern already established on other similar planets. Or perhaps ours is the first planet to follow this kind of developmental path, and there are others tagging along behind.

Evolving Habits of Life

Over a century ago, Samuel Butler pointed out that living organisms are essentially creatures of habit and suggested that they inherited an unconscious memory from their predecessors. The instincts of animals are the behavioral habits of the species. Likewise, the way in which organisms grow is habitual. As embryos develop, they pass through stages that recall the forms of remote ancestral types; in some way, the development of each individual organism seems to be related to the evolutionary process that gave rise to it. Human beings, for example, pass through a fish-like stage with structures like embryonic gills (Fig. 6.1). Butler saw in this a manifestation of the organism's inherited memory. "The small, structureless, impregnate ovum from which we have each of us sprung, has a potential recollection of all that has happened to each one of its ancestors."[14]

Such ideas were widely discussed by biologists until the early years of this century, and the idea that "heredity is a form of unconscious organic memory" was worked out in considerable detail.[15] By the 1920s, however, the development of genetics seemed to have shown that heredity could be explained in terms of genes and that it worked entirely mechanistically. Today, in the light of evolutionary cosmology and the possibility that all nature is essentially habitual, the idea of living organisms as creatures of habit takes on a new significance. Biological evolution may be not just a matter of material genes but of habits inherited nonmaterially.

How do new habits arise and evolve? In general, innovations occur in organisms in response to changes in the environment or in response to genetic mutations. Most innovations are not favored by natural selection; generally the habitual patterns of the species continue to predominate. The established "wild type" usually remains stable in most species for hundreds of thousands or even millions of years (in living fossils, such as the horsetail plants of the genus *Equisetum*, for 100 million years or more). In the language of genetics, most new mutant forms are recessive. In other words, when the mutant organisms are crossed with the normal or wild type, the habitual form of the species is dominant in the resulting hybrids.

When new mutant forms or new patterns of behavior are favored by natural selection, they become increasingly habitual, and they also tend to become increasingly dominant. Geneticists call this process the *evolution of dominance*. It is conventionally

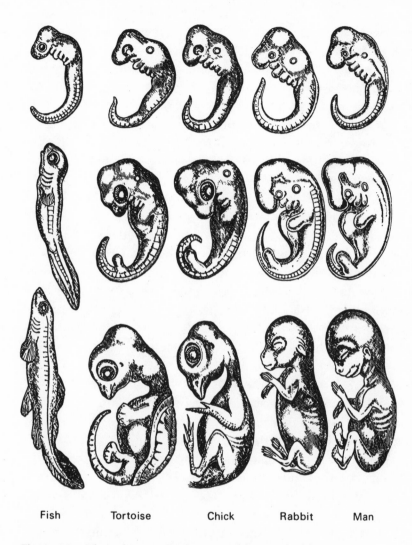

Fish   Tortoise   Chick   Rabbit   Man

Figure 6.1. The embryonic development of five species of vertebrates, illustrating the striking similarities at the early stages of development. Note the embryonic gill-arches between the eye and forelimb. (After Haeckel, 1892.)

explained in terms of hypothetical changes in the genetic structure of organisms, too subtle to be identified in detail.[16] But the idea of a new habit gradually building up by morphic resonance fits the facts just as well and provides an alternative explanation of why the wild type tends to predominate.[17]

We can also see the predominance of long-established habits in the way domesticated animals tend to revert to the wild type of their species when they escape from human influence. Feral cats, for example, start behaving like wild cats after only a short period in the wild. Many feral animals revert not only in behavior but in bodily form. Feral pigs, for example, become more bristly and tend to redevelop their tusks, and the stripes of young wild pigs reappear in their young. As Charles Darwin commented: "In this case, as in many others, we can only say that any change in the habits of life apparently favours a tendency, inherent or latent in the species, to return to their primitive state."[18]

Sometimes the habits of development of ancestral species, even species extinct for millions of years, reappear spontaneously in abnormal organisms, as when human babies are born with tails. Such freaks are called atavisms, reversions, or throwbacks. Occasionally these may be of considerable evolutionary significance. There are many examples from the fossil record that suggest that particular evolutionary pathways are repeated; organisms with features almost identical to previous species appear again and again. This process is called *evolutionary iteration*.[19]

Just as atavistic organisms may tune in to the habits of ancestral species by morphic resonance, other kinds of abnormality may involve picking up habits from contemporary species by morphic resonance, even species living on distant continents. Mutant organisms of this kind engage in a kind of evolutionary plagiarism, no doubt entirely unconsciously. Think, for example, of the remarkable instances of parallel evolution between placental mammals in the Old and New Worlds and marsupials in Australia, where similar forms of animals developed quite independently (Fig. 6.2).

Spontaneous Variation

If mutant organisms that have picked up some of the developmental or behavioral habits of other species are favored by natural selection, these features will become habitual by repetition and will become a normal aspect of this new kind of organism. Such unconscious borrowings may have played an important part in the evolutionary process. But evolution obviously involves more than the permutation and recombination of existing patterns of organization. Truly new patterns have to come into being for the first time—such as the first cell, eye, feather, spider's web, vertebrate, or bird—and each species represents a new variation on the general pattern of organization of its genus and family.

The conventional explanation of evolutionary creativity is in terms of random genetic mutations followed by natural selection. But this is more a dogmatic assertion than an established scientific fact. First, the idea that all mutations are random is only an assumption, and recent evidence from experiments with bacteria has thrown it into question. Some kinds of mutations seem to be purposive. For example, when starving bacteria are in the presence of a sugar they are constitutionally unable to use, genetic mutations occur at frequencies far above chance levels to give the bacteria particular enzymes they need, just when they need them.[20]

Secondly, the presence of a genetic mutation, even a truly random mutation, does not in itself explain how the organism adjusts and adapts to this genetic change. For example, if a bird is born blind as a result of a genetic defect that prevents it from making the visual pigment in the retina of the eye, it will not be able to develop the habitual instincts of its species that depend on vision. Most such birds probably perish young. Occasionally,

marsupials   placentals

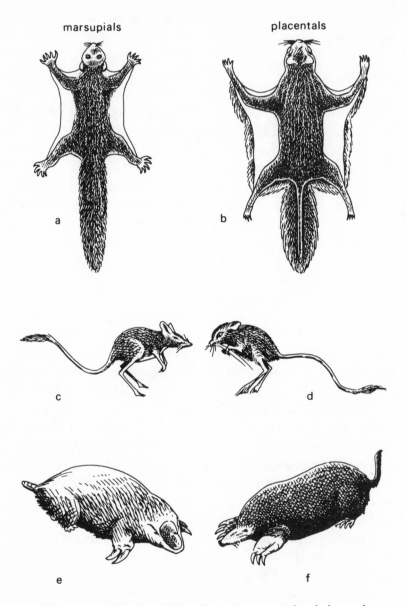

a   b

c   d

e   f

Figure 6.2. Examples of parallel evolution in marsupial and placental mammals. A and B: A marsupial flying phalanger and a placental flying squirrel. C and D: Marsupial and placental jerboas. E and F: Marsupial and placental moles. (After Hardy, 1965.)

by a creative leap, one such bird, while cheeping pathetically, might discover that it can find its way around by means of echoes. The development of such an echo-location pattern of behavior would be a creative response to its blindness, but would not be coded in the mutant gene that made it blind in the first place. If it was able to reproduce, its blind descendants would tend to develop the same echo-location pattern more readily by morphic resonance. The blind birds, able to move by echo location, might be at a considerable advantage at night and in dark caves. If favored by natural selection, they could give rise to a new species capable of living in such caves, with quite different behavioral habits from the parent species. Such birds actually exist: cave swiftlets, which live in dark caves in Asia, navigate in a similar way to that of bats.

Random mutations impose new necessities on organisms. They play a creative role insofar as necessity is the mother of invention. The way organisms adjust to the genetic mutation or environmental change may involve a creative leap, the synthesis of a new pattern of organization. On the hypothesis of formative causation, such patterns are organized by morphic fields, and these fields then grow stronger, becoming increasingly habitual, if favored by natural selection. Thus the creativity that gives rise to new bodily forms and to new patterns of behavior is not explained by the random mutations alone. It involves a creative response on the part of the organism itself and also depends on the ability of the organism to integrate this new pattern with the rest of its habits.

In Charles Darwin's evolutionary theory, the spontaneous variations of organisms were not ascribed to random genetic mutations: Darwin knew nothing of genes. He emphasized their dependence on the holistic coordinating powers of the organism, which he thought of in terms of the *nisus formativus*, the formative impulse postulated by early nineteenth-century vitalists. For example, "We may infer that, when any part or organ is either greatly increased in size or wholly suppressed through variation and continued selection, the coordinating power of the organization will continually tend to bring again all the parts into harmony with each other."[21]

Darwin took it for granted that acquired characteristics could be inherited and placed great emphasis on the role of habit in the evolutionary process. He provided many examples of the hereditary effects of the habits of life. For instance, in domesticated fowls, ducks, and geese, he noted the decrease in size of wing bones and increase in size of leg bones. "There can be no doubt that with our anciently domesticated animals, certain bones have increased or decreased in size and weight owing to increased or decreased use."[22] He thought similar principles applied under natural conditions; ostriches, for example, had lost the power of flight through disuse and gained stronger legs as a result of increased use over successive generations. And human evolution was no exception:

> Everyone knows that hard work thickens the epidermis on the hands; and when we hear that with infants, long before birth, the epidermis is thicker on the palms and the soles of the feet than on any other part of the body . . . we are naturally inclined to attribute this to the inherited effects of long-continued use or pressure.[23]

Darwin was very conscious of the power of habit, which for him was almost another name for nature. As he put it succinctly, nature made "habit omnipotent and its effects hereditary." Francis Huxley has summarized Darwin's attitude as follows:

> A structure to him meant a habit, and a habit implied not only an internal need but outer forces to which, for good or evil, the organism had to become habituated. . . . In one sense, therefore, he might well have called his book *The Origin of Habits* rather than *On the Origin of Species.*[24]

Thus the notion that the habits of nature evolve under the influence of natural selection is close in spirit to the thinking of Darwin himself, though at variance with the neo-Darwinian doctrines currently predominant in academic biology.

The Spread of New Habits

Creative responses to new environments and to new opportunities are a striking feature of plants and animals. Individual organisms have an inherent ability to adapt to their circumstances, within limits. For example, plants develop differently in different climates; they adapt to their environment. Animals can invent new ways of behaving and take advantage of new opportunities. As we have just seen, Darwin thought such acquired adaptations tended to become hereditary. The neo-Darwinian school rejects this aspect of Darwin's thought and denies that adaptations or learning acquired by individual organisms can be passed on to their descendants. According to neo-Darwinians, organisms just pass on the genes they themselves have inherited, and the only changes that occur in the genes are random. There can be no inheritance of acquired characteristics because there is no genetic mechanism by which such characteristics can be passed on.

However, if new habits inherited by morphic resonance build up within a species, new patterns of behavior *can* be passed on. Through repetition, there will be an increasing tendency for other members of the species to follow the same pattern of development or behavior under similar circumstances. The new behavior can spread by morphic resonance not just from parents to offspring but also to other members of the species in other places. There is already circumstantial evidence that this process actually takes place, in the realms of both morphogenesis and behavior.[25]

For example, around the beginning of this century, a system of milk delivery began in Britain in which bottles of milk were left every morning on doorsteps. The bottles were covered with cardboard caps. In the early 1920s, people in Southampton found that these caps were being torn off in shreds, and the top of the milk was being drunk. The culprits were birds called tits (closely related to the chickadees of North America). The habit spread locally, presumably through imitation. Tits are home-loving birds and do not usually venture more than a few miles from their breeding grounds; a movement of fifteen miles is exceptional. Nevertheless, this phenomenon soon started in other parts of Britain where it seemed to have been discovered independently. The spread of this cream-stealing habit was monitored systematically from 1930 to 1950. A detailed analysis of the records showed that the habit was discovered independently at least eighty-nine times in the British Isles and that as time went on, the rate of independent discovery accelerated.[26] So impressive was the effect that a leading British zoologist went so far as to suggest that something like telepathy might be involved.[27]

This cream-stealing habit also appeared in tits in Sweden, Denmark, and Holland. The Dutch records are particularly interesting. Milk deliveries ceased during the Second World War, and began again only in 1947–1948. The prewar tits that remembered the golden age of free cream would have died out by this time, but nevertheless attacks on bottles rapidly began again. The habit soon reestablished itself all over The Netherlands, and "it seems certain that the habit was started in many different places by many individuals."[28]

This may well be an example of morphic resonance at work in the evolution of behavior and illustrates how this process permits a far more rapid spread of a new habit than would be possible by means of random genetic mutations and genetic selection over many generations.

Creativity and Habit

In this chapter, I have been suggesting that the evolutionary process as a whole involves an interplay between creativity and habit. Without creativity, no new habits would come into being; all nature would follow repetitive patterns and behave as if it were governed by nonevolutionary laws. On the other hand, without the controlling influence of habit formation, creativity would lead to a chaotic process of change in which nothing ever stabilized.

Through evolutionary creativity, new patterns of organization come into being, and through repetition, these new patterns become increasingly habitual if favored by natural selection. The already-established habits of nature, culture, and mind provide the context in which further creativity occurs and in which new patterns are subject to natural selection.

Obviously the notion that the regularities of nature, culture, and mind are habitual cannot explain how new patterns arise in the first place: it cannot account for creativity, only for the stabilization of new patterns once they have come into being. What are the sources of evolutionary creativity? I return to this question in Chapter 9.

PART THREE

THE
REVIVAL
OF
ANIMISM

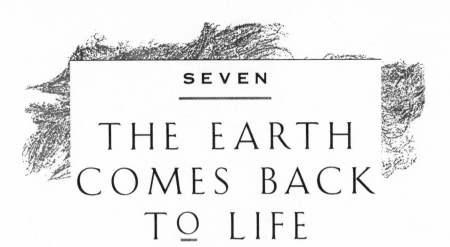

SEVEN

THE EARTH COMES BACK TO LIFE

The Rediscovery of Mother Earth

In the last few centuries, an educated minority in the West has believed that our planet is dead, just a misty sphere of inanimate rock hurtling around the sun in accordance with mechanical laws. This is a very eccentric opinion when seen in a larger human context. Throughout history, practically all humanity (and the majority even today) have taken it for granted that the earth is alive.

The emerging contemporary understanding of the earth as alive, though rooted in old patterns of mythic thought, is strongly influenced by two characteristically modern perceptions: first, the vision of the earth from space, as seen by astronauts and cosmonauts; second, the realization that our economic activities are changing the global climate.

The view of the earth as a spherical body floating in space and

rotating on its axis was grasped imaginatively at the very beginning of the scientific revolution. A commonplace version of this
idea has been familiar to generations of schoolchildren through
the use of globes. Perhaps the inanimate quality of these simple
models implicitly reinforced the idea that the earth itself was
inanimate. At any rate, the view of Earth from orbiting artificial
satellites and from the surface of the moon in one sense simply
confirmed what most educated people already believed. This vision was a triumph for humanity not only because of the space
explorers' heroism and the technology that made their journeys
possible but also because it confirmed the power of scientific
imagination. This is how it struck the Russian cosmonaut Igor
Volk:

> Several days after looking at the Earth a childish thought
> occurred to me—that we cosmonauts are being deceived.
> If we are the first ones in space, then who was it who made
> the globe correctly? Then this thought was replaced by
> pride in the human capacity to see with our mind.[1]

At the same time, this vision of the earth from outside had a
deeper and more mystical impact. Many of the space explorers
were moved by her beauty, her purity and splendor. Others were
fascinated by her ever-changing aspects, impossible to capture
in static photographic images. "The clouds were always different, the light was different. Snow would fall, rain would fall—
you could never depend on freezing any image in your mind."[2]

But the cosmonaut Aleksandr Aleksandrov summed up the
principal message for millions of people. Looking down on
America and then on Russia, he saw the first snow and imagined
people in both countries getting ready for winter. "And then it
struck me that we are all children of our Earth. It does not matter what country you look at. We are all Earth's children, and
we should treat her as our Mother."[3]

The realization that we are polluting the earth, upsetting the balance of nature, and changing the global climate points to the same conclusion. The destructive forces unleashed through economic development and the growth of technology have taken on a life of their own, proceeding in blithe disregard of their planetary consequences. And they have been accompanied by unprecedented growth of the human population. These processes now seem unstoppable. But our activities are not separate from the earth. We live within her. If we disregard her in pursuit of our human ends, we endanger our own survival. Like the Great Mother of ancient mythology, she has a terrifying aspect:

> Gaia, as I see her, is no doting mother tolerant of misdemeanours, nor is she some fragile and delicate damsel in danger from brutal mankind. She is stern and tough, always keeping the world warm and comfortable for those who obey the rules, but ruthless in her destruction of those who transgress. (James Lovelock, 1988)[4]

Our human dependence on the living processes of the earth was largely forgotten with the growth of industrial civilization. Now we are being forced to remember that Gaia is greater than we are and that the human economy is embedded within the ecology of the biosphere. So, in what sense is Gaia alive? And what difference does it make if we think of her as a living organism, as opposed to an inanimate physical system?

The Life of the Earth

The organismic or holistic philosophy of nature that has grown up over the last sixty years is a new form of animism. It implicitly or explicitly regards all nature as alive (Chapter 5). The universe as a whole is a developing organism, and so are the galaxies, solar systems, and biospheres within it, including the earth (Chapter 6).

From the mechanistic point of view, these ideas are nonsensical. There is no such thing as "life," just complex patterns of mechanistic interaction taking place in accordance with the eternal laws of physics and chemistry. Biological organisms are complex mechanisms that have evolved through random genetic mutations and natural selection. Since the earth does not reproduce, does not have genes, and has not (as far as we know) evolved as a result of competition and natural selection, it is not alive. Even biological organisms, such as dolphins and bamboos, which we conventionally refer to as living, are not alive in the sense that they are animate. They are just complex self-regulating mechanisms.

Not surprisingly, it is practically impossible to define life in mechanistic terms. An inherent life or vitality in living organisms is just what the dreaded heresy of vitalism has always affirmed and what mechanism has always denied (Chapter 5). So how can the living state be distinguished from the nonliving? If an attempt is made to define life in molecular terms—for example, in terms of the possession of DNA and proteins—it becomes difficult to distinguish living from recently dead organisms since they both possess the same chemicals. Moreover, such chemical definitions would rule out the possibility of biological life elsewhere in the universe built on any other chemical basis. And this would seem too rash a generalization.

If life is defined in terms of physicochemical *processes,* it is again hard to say why such processes are features of life as opposed to equally mechanistic physicochemical processes in dead organisms or machines. Then what about reproduction, or at least the capacity to reproduce, as the essential feature of living organisms? This won't do either: Think, for example, of sterile organisms such as mules and worker bees. In the face of such difficulties, the question of the nature of life is usually ignored in academic science. If you try to look up *life* in most dictionaries of biology, you will find that it is conspicuous by its absence.

Some of the most promising modern attempts to conceive of the nature of living organisms involve concepts of information, communication, and control, including an emphasis on the importance of feedback. This is the approach of systems theory and the holistic or organismic philosophy in general. Organisms are living wholes, self-organizing processes of activity. Biological organisms are just one kind of organism. The earth is a far vaster organism, within which they come into being, develop, reproduce perhaps, and sooner or later die. She is indeed far more like a great Mother than a misty ball of inanimate rock.

Arguments between mechanists and holists have raged for decades. What is at stake are fundamental models or paradigms of reality. In the context of growing fears of environmental crisis, it has become obvious that our attitudes affect the way we live and even affect our prospects of survival as a species. The debate over mechanistic and animistic models of reality is now not just a scientific or philosophical issue but also a political one.

The Gaia Hypothesis

James Lovelock, the leading proponent of the hypothesis that Earth is a self-regulating living organism, first began to formulate his ideas when he was thinking about ways of detecting life on Mars. He realized that if Earth's atmosphere were made up of gases in chemical equilibrium, like the atmospheres of Mars and Venus, it would consist of about 99 percent carbon dioxide. Instead, it contains 0.03 percent carbon dioxide, 78 percent nitrogen, and 21 percent oxygen. This composition could have come about only through the activities of living organisms and could be maintained only through their continuing activity.

It is now generally accepted that there was virtually no oxygen and nitrogen in the earth's atmosphere to start with; their current predominance is due to the nitrogen-releasing activities of bacteria and the evolution of photosynthesis, releasing free oxygen. The reduction of carbon dioxide to its present low levels is likewise due to biological activity, through which vast quantities of carbon have been removed from the atmosphere and buried—

for example, as calcium carbonate in limestone rocks. (These are largely made up of the shells of tiny organisms in the oceanic plankton, which were deposited as sediments on the sea floor.)

Lovelock points out that the atmosphere, the weathering of rocks, the chemistry of the oceans, and the geological structure of the earth have been so profoundly modified by biological activities that all these interlocking systems can be understood only in relation to one another. Together, they interact to maintain a remarkable and long-lasting stability without which the evolution and continued existence of living organisms would not be possible. They constitute a single living system, which can be called the biosphere, or Gaia. Gaia is "a self-regulating entity with the capacity to keep our planet healthy by controlling the chemical and physical environment."[5] One of the regulatory activities of the biosphere is to maintain the planetary temperature within the narrow limits necessary for biological life. In this process, the concentration of "greenhouse gases" such as carbon dioxide in the upper atmosphere plays a major role.

To take another example of Gaian self-regulation, the concentration of salt in the seas has been maintained at a similar level ever since the oceans first came into being and life began about 3,500 million years ago. If the salt concentration was much higher than the present level, around 3.5 percent, life in the oceans would be impossible. Yet salts are being added to the sea all the time. Some come from the weathering of rocks on land, and are washed into the sea through rivers; others well up from the interior of the earth as hot rocks push through the ocean floor (as they do in the sea-floor trenches between the great continental plates). At the rate salts are being added to the sea today, it would have taken less than 80 million years to reach the present levels.[6]

Obviously, the maintenance of salts at more or less steady levels over much longer periods means that they must somehow be removed at a similar rate. One way this can happen is through the formation of lagoons that get separated from the sea and evaporate, leaving salt deposits that are then buried. Just how the rate of lagoon formation is regulated is unknown, but Lovelock suggests that the formation of rocky reefs by colonies of microorganisms in shallow waters could have played a part by forming the necessary barriers. The movements of continental plates, leading to the formation of mountains and the folding of rocks at continental margins, must also have had a major influence. He points out that movements of the earth's crust may themselves be influenced by biological activity through the sheer mass of the limestone rocks deposited on the sea floor.[7]

Physiology is the study of the processes that go on in animals and plants and the interrelations between these processes that enable organisms to maintain a more or less steady state. Lovelock calls the analogous science of the living earth *geophysiology*.[8] This science focuses attention on the processes of planetary regulation, and brings together areas of inquiry that are normally pursued in separate academic disciplines: geology, geophysics, oceanography, climatology, ecology, biology, and so on. Such a planetary perspective is clearly essential if we are to understand the evolutionary history of Gaia, including the way the biosphere has been able to survive the progressive heating-up of the sun and recover from catastrophic events such as ice ages and the impact of asteroids. Such collisions of massive bodies with the earth probably caused the series of sudden mass extinctions revealed by the fossil record. The last of these global disasters occurred some 60 million years ago when dinosaurs and many other forms of life suddenly died out.

But geophysiology is not just of academic interest; it is of immediate practical relevance. No one knows what effect the increasing concentrations of greenhouse gases in the atmosphere will have on the global climate. Changes may be gradual, but there may also be relatively sudden changes; for example, in the pattern of circulation of the great oceanic currents such as the Gulf Stream, with unknown climatic consequences. Likewise, no one knows the likely consequences of the destruction of tropical rain forests, which through transpiration and the encouragement of cloud formation have a powerful cooling effect. Desert formation on a huge scale is one probable consequence, but there could well be more wide-ranging effects on the global climate. And no one knows how pollution of the seas will affect the activity of plankton, which play such a major role in the regulation of the chemical composition of the atmosphere.

The development of Gaian science may not enable the consequences of these changes to be predicted very accurately, but at the very least it makes us more aware of some of the potential effects of human economic activity, not to mention nuclear war. It gives a broader perspective than those provided by the annual balance sheets of businesses, annual economic growth rates, and the short-term problems of politics. There is no mistaking the need for such a perspective as we actually experience the climate changing around us.

The idea of Gaia is deeply upsetting to both mechanistic science and humanism.[9] Mechanists reject the idea of Gaia as alive, which is not surprising in view of their traditional objections to the vitalist proposition that plants and animals are animate. Some defenders of the Gaia hypothesis prefer to try to stay within the conventional framework of scientific thought, concentrating on physical and chemical interactions discussed in entirely mechanistic terms, with no hint of mysterious vital properties or hidden purposes. The microbiologist Lynn Margulis, for example, a close associate of Jim Lovelock's in the initial formulation of the Gaia hypothesis, disavows the more radical aspects of his thesis:

> I reject Jim's statement "The Earth is alive"; this metaphor, stated this way, alienates precisely those scientists who should be working in a Gaian context. I do not agree with the formulation that says "Gaia is an organism." First of all in this context no-one has defined "organism." Furthermore I do not think that Gaia is a singularity.

Rather Gaia is an extremely complex system with identifiable regulatory properties which are very specific to the lower atmosphere.[10]

In response to the criticism that his thesis implies that Gaia is purposive, Lovelock himself has put forward computer models that show that some regulatory processes could occur by purely mechanistic means, involving no more than the operation of the laws of physics and chemistry. But still the fundamental problem will not go away. Conventional physiologists try to explain the functioning of plants and animals in mechanistic terms, without invoking any animating principles or purposes; nevertheless, morphogenesis, instinctive behavior, learning, and memory are still among the unsolved problems of biology, and the nature of life itself remains an open question (Chapter 5). Even if some aspects of Gaian self-regulation can be modeled mechanistically, the nature of Gaia's life likewise remains a mystery.

The Gaia hypothesis is undoubtedly a major step in the direction of a new animism, which is why it is so controversial. Its widespread appeal depends on the way in which it reconnects us with premechanistic and prehumanistic patterns of thought. Many scientists may prefer a watered-down version of the type advocated by Lynn Margulis, but this cannot disguise the fact that it involves a radical shift in point of view from the man-centered world of humanism to a recognition that we all depend on the providence of Gaia. As Lovelock puts it:

> Gaia theory is as out of tune with the broader humanist world as it is with established science. In Gaia we are just another species, neither the owners nor the stewards of this planet. Our future depends much more upon a right relationship with Gaia than with the never-ending drama of human interest.[11]

The Purposive Development of Gaia

If Gaia is in some sense animate, then she must have something like a soul, an organizing principle with its own ends or purposes. But we need not assume that the earth is conscious just because she seems to be alive and purposive. She *may* be conscious, but if so, her consciousness is likely to be unimaginably different from our own, which is inevitably shaped by human culture and language. On the other hand, she may be entirely unconscious. Or she may, like ourselves, be a creature of unconscious habit with some degree of consciousness some of the time. This question has to be left open.

The conscious or unconscious purposes of Gaia include the development and maintenance of the biosphere, and they must in some sense include the evolution of humanity. Just as the cosmological anthropic principle can be stated in weak and strong forms, so can the Gaia hypothesis. The weak form amounts to saying that Gaia must be constituted in such a way as to have permitted the survival and evolution of life for billions of years; otherwise, we would not be here to discuss it. The weak form is indeed weak because it can always be construed by its opponents as "a slightly supernatural restatement of the fact that life on Earth has survived, at least so far."[12] The strong form recognizes that Gaia herself is purposeful and that her purposes are reflected in the evolutionary process. It raises the difficult question of the nature of the purposive organizing principle, traditionally regarded as the soul or spirit of the earth.

In modern evolutionary physics, the old idea of the soul of the universe has been replaced by the idea of the primal unified field, from which the known fields of physics arose, and of which they are aspects. Likewise, the soul of the earth may best be thought of in terms of the unified field of Gaia. Her gravitational and electromagnetic fields are aspects of this field but not its only aspects. We may recall that the founder of the modern science of magnetism, William Gilbert, thought of the earth's magnetic power in terms of its soul and tried to explain gravitation in similarly animistic terms.

The earth's gravitational field extends all around her, causing the moon to orbit the Earth and relating her to the gravitational fields of the sun and other bodies in the solar system. Her magnetic field is more local, but nevertheless extends far beyond her surface (Fig. 7.1).[13] This field is remarkably variable and has changed considerably over the last few centuries (Fig. 7.2). Moreover, the north and south magnetic poles are continually wandering, and on a geological time scale, the magnetic poles reverse quite frequently. In the last 20 million years, for example, the magnetic North Pole has been at the geographic South Pole over forty times and has remained there for periods of up to a million years.[14] (The history of these polar reversals has been reconstructed from the direction of magnetization in rocks; these provide a fossil record of the magnetic polarity prevailing

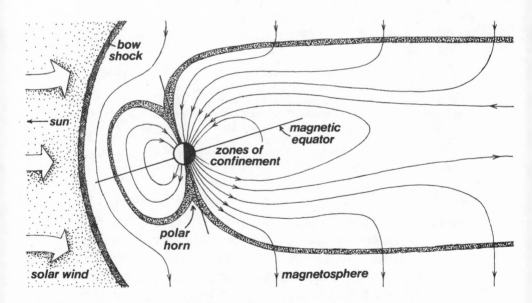

Figure 7.1. The earth's magnetic field, confined by the solar wind into a comet-shaped cavity called the magnetosphere. The wind compresses the magnetosphere on the day side to a distance of about 10 earth radii. On the night side, the wind sweeps the earth's magnetic field into the "magnetotail" which extends for at least 1,000 earth radii.

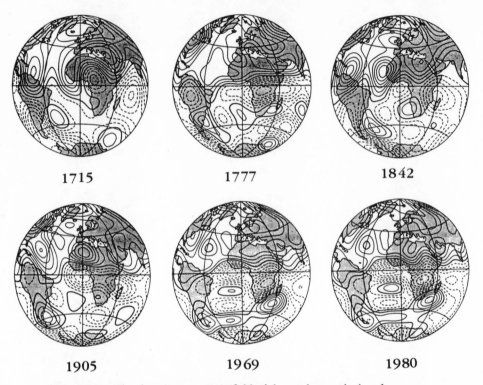

1715   1777   1842

1905   1969   1980

Figure 7.2.   The changing magnetic field of the earth over the last few
centuries. The contour plots indicate the strength of the field at the boundary
between the molten core and the mantle. (The lines of force come out of the
Southern and flow back into the Northern Hemisphere, as shown in Figure
7.1. The solid contours represent the intensity of magnetic flux into the core;
broken lines flux out of the core. (From Bloxham and Gubbins, 1985.)

at the time they formed. A reversal of polarity is indicated by
the reversed magnetization of successive layers of rock.)

No one knows why or how these reversals of polarity have
occurred, apart from supposing that they must have depended
on changes in the circulatory patterns of molten rock in the interior of the earth, which act rather like a dynamo. Such patterns
of circulation also power the spreading of the sea floor between
the continental plates and thus shape the morphogenesis of the
continents and oceans. These unknown processes in the unknown heart of Gaia somehow control the development of the

surface of the earth and the magnetic field within and around it. Are these changes just a matter of chance? Or is there some deeper organizing principle behind them?

Take a specific question about Gaian morphogenesis, which has intrigued me for years. Why are the geographic poles polar? They are polar opposites in the sense that at the North Pole is an ocean surrounded by continents; at the South Pole, a continent surrounded by oceans. Is this merely a chance coincidence, a fortuitous result of aimless continental drift? Or could this pattern be the culmination of a morphogenetic process? Such a morphological polarization of a spherical body is very familiar in the realm of biology; for example, in the formation of poles in fertilized eggs. In plant eggs, the primary polarity is between the shoot and root poles; in animal eggs, between the so-called animal and vegetal poles, the animal pole being nearest the nucleus of the egg, the vegetal pole containing the nutritive yolk.

If we think of the earth in terms of biological analogies, then she is more like a developing organism than a stable fully grown adult. In recent geological time there has been a series of ice ages, and in earlier times too the climate and conditions of life were continually changing. Some of these changes may have been triggered by variations in the sun's activity, both over the long term as the maturing sun grows hotter and through more short-term cycles of activity within it. The sun's magnetic field, for example, shows a complex resonant pattern of cycles, including a reversal of polarity about every eleven years, coinciding with peaks of sunspot activity as great flares of incandescent gas shoot far out from its surface.[15] The last such reversal occurred in early 1990.

Other changes in Gaia's conditions of life may have resulted from the movements of the solar system in relation to the rest of the galaxy, with its 34-million-year period of oscillation up and down through the plane of the galactic disk, and its 285-million-year orbit around the center of the Milky Way. These cycles may be related to periods of reversal of the earth's magnetic field and also to mass extinctions.[16] Moreover, sudden catastrophic changes have probably been caused on dozens of occasions by the collision of massive bodies (such as asteroids) with the earth.

But although the development of Gaia is subject to a variety of external influences, she may also have her own internal purposes, just as the growth of an embryo, though perturbed by external influences, follows a pathway that leads it toward the mature form of its species. As far as the embryo is concerned, this goal lies in the future. In the language of dynamics, the developing embryo moves toward its morphogenetic attractor. This attractor is contained in the organism's morphogenetic field.

Is the evolution of Gaia moving toward an attractor? And if so, what part is humanity playing in this developmental process? Obviously, answers to these questions must be speculative. But if the universe itself is evolving in a purposeful way, it makes sense to ask the same questions about Gaia, even if her goals are obscure and can be inferred only from what we know of the earth's evolutionary development so far.

The Morphic Field of Gaia

In terms of the hypothesis of formative causation, the purposive organizing field of Gaia can be thought of as her morphic field. Such fields animate organisms at all levels of complexity, from galaxies to giraffes, from ants to atoms. They organize, integrate, and coordinate the constituent parts of organisms so that the whole system develops in accordance with its characteristic ends or goals; they maintain the integrity of the system and enable it to regenerate after damage. Thus, in terms of these general principles, we would expect the morphic field of Gaia to coordinate her various constituent processes, such as the circulation of molten rocks in the interior, the dynamics of the magnetosphere, the movements of continental plates, the circulatory patterns of the oceans and the atmosphere and their chemical composition, the regulation of global temperature, and the evolution of ecosystems. These regulatory activities, like those of morphic fields in systems at all other levels of complexity, would involve an ordering of otherwise indeterminate, probabilistic processes.

The idea of the morphic field of Gaia might at first glance seem just to attach a new name to the soul of the earth or to mean no more than vague terms such as "complex self-organizing system," "self-regulatory properties," or "emergent holistic principles." But I believe it does much more than this. First, it places Gaia within the framework of a testable hypothesis that applies throughout the realm of nature. This means that empirical research on the morphic fields of chemical and biological organisms could reveal general features of such fields and thus indirectly deepen our understanding of the morphic field of Gaia. Second, it raises the possibility that Gaia could be interacting by morphic resonance with other Gaialike planets elsewhere in the universe (p. 135). And third, it implies that through self-resonance, the morphic field of Gaia contains an inherent memory. Like other organisms, Gaia builds up habits through repetitive patterns of activity. The more often these patterns are repeated, the greater the likelihood of their happening again.

We know very little about Gaia and her purposes. We know very little about morphic fields. But there is hope of finding out more. And in the meantime, whatever theories we may have of the nature of Gaia, at least we have to recognize that our own lives depend on hers. If we take her for granted, we do so at our peril.

EIGHT

SACRED
TIMES AND
PLACES

Space, Place, and Time

We live in particular places; things happen at particular times. The different qualities of places and times are apparent to everyone. Think, for example, of the contrasts between our homes and places of work, cities and countryside, day and night, winter and summer, Christmas and Easter. These all have their own qualities; they tend to induce different emotions, feelings, habitual attitudes, and states of mind.

Our experience of the qualities of places and times is obviously shaped by our own emotions and preoccupations, by our personal memories and our biological, cultural, and religious heritage. But it is also influenced by the places and times themselves. Lived experiences involve a combination of all these factors. Our experience of our environment is not "objective" in the sense that it is a mechanical response to the immediate physical conditions, measurable by scientific instruments. It has a social, cultural, and religious dimension, as well as a unique personal aspect. According to the hypothesis of formative causation, the conscious and unconscious memory of places and times is strongly influenced by morphic resonance.

Mechanistic science has very little to tell us about the qualities of times and places. As a matter of principle, in the name of scientific objectivity, the observer's subjective responses are supposed to be stripped away and ignored (Chapter 2). The human element is eliminated to produce a model of reality that contains only measurable quantities that can be related to one another through mathematics. The behavior and properties of these abstracted quantities—such as mass, momentum, electric charge, and temperature—are supposed to follow eternal laws that are the same at all times and in all places. For the same reasons, all scientific experiments are supposed to be exactly repeatable anywhere in the world at any time, given the same physical conditions. It should not matter whether they are performed in Berkeley or Bali, Cambridge or the Cameroons, nor should it matter when they are done. In practice, of course, scientific experiments tend to be performed in a few major centers of research, places with their own qualities and traditions, and they are usually performed in the daytime on weekdays.

Insofar as experiments *are* repeatable, they are so because the experimental system was isolated as far as possible from the particular qualities of the environment around it; for example, the experiment may have been done in artificial light in a thermostatically controlled environment. Such experimental techniques are suitable for studying the lowest common denominators of physical and biological processes. But they obviously isolate and abstract only certain aspects of these processes, ignoring all those that cannot be measured or controlled by the researcher.

Of course, actual experience, not the limited abstractions of science, matters most in the conduct of our lives. And it is our entire experience, including our cultural heritage, that links us to the world in which we live, not just the artificially limited aspects of experience that constitute an experiment or a scientific observation. If we are not to live double lives, split between an "objective" impersonal mechanistic reality and the "subjective" world of personal experience, we need to find a way of bridging these two realms.

Mechanistic science cannot guide us in this endeavor because it depends on creating this split in the first place. By contrast, an evolutionary, holistic science of the future should be able to help in this process of integration. The Gaia hypothesis is already a step in this direction. In this chapter, I consider the nature of the integrative process in the practices of traditional societies, from which we may have much to learn.

Seasonal Festivals

The most obvious way the life of human communities is linked to the earth and the heavens is through the seasonal festivals found in societies all over the world. These festivals are related to the cycles of the sun, vegetation, and animal life. Even the lives of modern city dwellers are still ordered in annual cycles by the major traditional holy days.

For pastoralists, hunters, and farmers, the annual patterns of human activity are deeply intertwined with the seasons, the movements of animals, and the growth of plants. They are part of the seasonal cycles of nature, not separate from them. The seasonal festivals express in ceremonial form this participation of the whole social group in the rhythms of the living world.

In pre-Christian Europe, the four great solar festivals were held around the shortest and longest days, the midwinter and midsummer solstices, and around the times when day and night are of equal length, the spring and autumn equinoxes. All these sacred times were assimilated into Christianity: the midsummer festival, St. John's Day; the spring festival, Easter, which is a movable feast because it depends on both sun and moon (Easter Sunday is always the first Sunday after the full moon that happens upon or next after the vernal equinox); and soon after the autumn equinox is Michaelmas, the feast of St. Michael and All Angels; and of course the midwinter festival became Christmas.

In northern Europe and North America, Christmas has taken on a wide variety of ancient symbolic associations. The birth of the sacred child is linked to the rebirth of the solar year as the days begin to lengthen again. In the erecting and decorating of the Christmas tree is a contemporary veneration of the sacred cosmic tree, the enduring source of life and renewal. Santa Claus can even be seen as a visiting shaman from the frozen North—indeed from Lapland, the last outpost of shamanism in Europe—who flies through the air with his reindeer.[1]

Between these four solar festivals were the four fire festivals, at the beginning of November, February, May, and August. The festival at the beginning of November was the Celtic new year, the time when the spirits of the dead returned as the old year gave way to the new. This became the great Christian festival of the dead, still celebrated as All Saints Day on November 1 and All Souls Day on November 2. In England, the old custom of burning a human image of the old year in the form of a man (linked to the tradition of human sacrifice by fire) survives in the form of Guy Fawkes Day on November 5. The eve of the feast of the dead, Halloween, is still celebrated in North America and elsewhere with turnips or pumpkins hollowed out like skulls, and the mischievous spirit of "trick or treat" is a dim echo of the orgies and reversals of the social order found all over the world at turning points in the year, representing a brief return to the primal chaos from which the order of the cosmos is created anew.

The fire festival at the beginning of February was Christianized as Candlemas on February 2; the August festival as Lammas on August 1 when the first fruits of the harvest were sanctified. But May Day remained the festival of the pagan goddess of fertility, Maia, after whom the month is named. In seventeenth-century England, these festivities scandalized the Puritans, who did their best to suppress them:

> Young men and maids, old men and wives, run gadding overnight to the woods, groves, hills and mountains where they spend all night in pleasant pastimes; and in the morning they return, bringing with them birch and branches of trees, to deck their assemblies withal. . . . But the chiefest jewel they bring from thence is their May-pole, which they bring home with great veneration. . . . [It] is covered all over with flowers and herbs, bound around with strings from the top to the bottom, and sometimes painted with variable colours. . . . And this being reared up, with handkerchieves and flags hovering on the top, they strew the ground round about, bind green boughs about it, set up summer halls, bowers and arbours hard by it. And they fall to dance around it, like as the Heathen people did at the dedication of Idols, whereof this is the perfect pattern, or rather the thing itself.[2]

The themes of the great annual festivals are universal—the death of the old, the birth of the new, the fertility of people, animals, plants, and the earth. Even today, in the secularized world of the West, the ancient symbols retain something of their power. At Easter the rites commemorating the sacrifice of Jesus on the tree, his burial and resurrection, recall the dying and resurrected fertility gods of the ancient world, such as Attis and Osiris, as well as the annual sacrifices of sacred kings or other human victims to ensure the fertility of the land.[3] And associated with the Easter festivities are other ancient symbols of fertility, such as rabbits, eggs, and chicks.

In the urban industrial West, such seasonal festivals still remind us of our communal participation in the cycles of nature. Through entering into the spirit of these festivals, we can link the material aspects of our lives with their social, mythic, and spiritual aspects. Thus, for example, the American Thanksgiving dinner is not just a delicious feast of roast turkey, pumpkin pie, and so on; just a social festival when families and friends come together; just a ritual remembering of the heroic founding fathers and mothers; just a thanksgiving for the abundance of the earth and the grace of God. It combines all these elements, and moreover, its ritual quality, rooted in tradition, relates modern Americans to all those who have celebrated this festival before them; it helps to define their identity as Americans.

Rituals and Morphic Resonance

All societies have their rituals, such as those of the seasonal festivals, the rituals of birth, marriage, and death, and the rituals that commemorate and reenact original events, charged with spiritual power, on which the social and religious group depends. For example, the Jewish feast of Passover recalls the original Passover dinner on the night when the firstborn of the Egyptians and their cattle were destroyed, after which the Jews began their journey out of bondage in Egypt. The Christian celebration of the Eucharist commemorates the last supper of Jesus with his disciples, itself a Passover dinner. The American national festival of Thanksgiving recalls the first thanksgiving dinner of the Pilgrims after their first harvests in the New World.

A general feature of all rituals is that they are intensely conservative. In order to work properly, they are supposed to be performed in the correct and customary manner. In many parts of the world, the very language of ritual is archaic, preserving the traditional form of words believed to be necessary for their efficacy. The liturgy of the Coptic church in Egypt is still carried out in the ancient Egyptian tongue; the Brahminic rituals of India, in Sanskrit.

Through this ritual participation, the past becomes present. The present participants are linked to all those who have gone before—to the ancestors, and ultimately to the primal creative moment the ritual commemorates. In Christianity, for example, this is the basis of the doctrine of the Communion of Saints. The sacred time of the mass is not only linked with that of preceding and following masses;

> it can also be looked upon as a continuation of all the Masses which have taken place from the moment when the mystery of transubstantiation was first established until the present moment. . . . What is true of time in Christian worship is equally true of time in all religions, in magic, in myth and in legend. A ritual does not merely repeat the ritual that came before it (itself a repetition of an archetype), but is linked to it and continues it, whether at fixed periods or otherwise. (Mircea Eliade, 1958)[4]

Why are rituals so conservative? And why do people all over the world believe that through ritual activities they are participating in a process that takes them out of ordinary secular time and somehow brings the past into the present? The idea of morphic resonance provides a natural answer to these questions. Through morphic resonance, ritual really can bring the past into the present. The present performers of the ritual indeed connect with those in the past. The greater the similarity between the way the ritual is performed now and the way it was performed before, the stronger the resonant connection between the past and present participants.

The Changing Qualities of Time

As well as the cyclical changes of day and night, of lunar months and the seasons, we are subject to rhythms of social and cultural activity that arise from the way our clocks and calendars are structured, for example working from nine A.M. to five P.M., and working on weekdays but not weekends. Then there are historical periodicities. Many countries celebrate particular days that were turning points in their political history—Independence Day, Republic Day, and so on—that take on something of the quality of seasonal festivals. And then there are longer-term celebrations, such as centenaries. For example, 1992 has a particular significance as the five-hundredth anniversary of the "discovery" of America by Christopher Columbus, and the celebrations will recall its momentous historical consequences. On a larger scale, the approaching dawn of a new millennium in the Christian era will inevitably create a sense of the beginning of a new age and the end of an old.

These celebrations remind us that we live in historical time, not just in a world of cyclical recurrences. Historical time has a cumulative quality, expressed in the very way the years are counted from the birth of an era: the Roman calendar was dated from the founding of the city by the eponymous Romulus; the Jewish calendar, from the supposed date of the Creation, 3,760 years before the beginning of the Christian era; the Christian era, from the birth of Christ; the Moslem era, from the Prophet's flight from Mecca to Medina in 622 A.D., and so on. This process is analogous to the way a person's age is expressed and implies a historical process of growth and maturation. Each era was once young, and it grows older year by year, just like a tree, an animal, or a person. And indeed the historical eras of civilization are traditionally thought of in terms of birth, youth, maturity, and senescence: empires rise and fall. This idea still retains its ancient plausibility. Is the civilization of the West now entering a phase of decadence? Will the rising economic and political power of Japan overtake it in global influence? We hear such questions discussed continually, and they have a great effect on our individual and collective sense of time. For example, in Britain, as a declining postimperial power, there is a chronic nostalgia; in the rising industrial powers of Asia, a pervasive optimism and sense of opportunity.

The evolutionary theory of life, together with the scale of geological time, divided into eras and periods, extends this sense of historical development to the entire biosphere. Gaia herself is developing, and the quality of time today is very different from that of the Precambrian (the age of microbes), or the Cretaceous (the age of dinosaurs); what can happen now is very different from what could happen then. Since the 1960s, the sense of historical time has been extended to the entire cosmos, a vast developing organism about fifteen billion years old, still growing and developing.

In general, the quality of time as we experience it depends on the development of the greater organized systems within which we live. There is no such thing as a featureless background mathematical time, flowing on uniformly forever. Time arises *within* developing systems. An embryo, for example, by its very nature is developing toward its future mature form; it has its own arrow of time and passes through a series of stages, each with its own quality. In a similar way, time is internal to the cosmos as a whole, and internal to all developing systems within it.

The idea of the quality of time is fundamental to astrology, which attempts to characterize the changing qualities of time through the relative positions of the celestial bodies. The quality of time at birth and at the inception of any new enterprise is believed to be of particular importance for its subsequent development. In India, for instance, the times and dates of weddings and other important events are still generally chosen on astrological grounds. The sense that some times are more auspicious for particular activities than others is deep-rooted. And whatever the predictive value of astrology (about which I for one am doubtful), it seems entirely reasonable to suppose that events on earth are related to their changing astronomical environment.

Finally, we all experience the quality of time through what Germans call the *Zeitgeist,* the "spirit of the time." No one knows why different periods of human history should have particular moods, feelings, and fashions; the spirit of the late 1960s, for example, felt very different from the spirit of the Reagan years. The practice of science is not immune from such changes in the cultural climate; like all other social activities, it is colored by them and in turn contributes to them.

The zeitgeist is not just generated within particular cultures and economic systems but depends on their interrelations with one another and with the broader natural environment on earth and in the heavens. Indeed, in the context of evolutionary cosmology, we can generalize the idea of the zeitgeist and see that it has a cosmic dimension. The universal spirit of the time was very different in the first few seconds after the Big Bang than it was when galaxies began to form, and it was different again when planets came into being. The changing times permitted new forms of evolutionary creativity to occur, new forms and patterns of activity to be repeated, and these in turn changed the quality of time. Ultimately the changing cosmic zeitgeist depends on the continuing growth of the universe, which gives a direction, an arrow of time, to the entire evolutionary process.

The Spirits of Places

> Different places on the face of the earth have different vital effluence, different vibration, different chemical exhalation, different polarity with different stars: call it what you like. But the spirit of place is a great reality. (D.H. Lawrence)[5]

The qualities of places are traditionally conceived of in terms of the *genius loci,* the "spirit of the place." In this context, the word *spirit* has two connected meanings: a feeling, atmosphere, or character; and an invisible entity or being, with its own soul and personality. It is difficult to disentangle these meanings, for the second could be thought of as a personification of the first. But then some people claim to experience the presence of beings in particular places. Are these simply psychological projections?

Or are they an intuitive way of relating to the living quality of the place, which may indeed have a kind of personality?[6]

Places traditionally associated with the presence of nature spirits are not distributed equally across the landscape. They are concentrated in particular areas, such as waterfalls, springs, streams, and rivers; in and around various trees; in caves and grottoes; and in parts of woodland, desert, moorland, mountains, and seashore. The nature spirits of such places were given generic names in classical mythology: naiads were water spirits; dryads, the spirits of trees and woodland; oreads, mountain spirits; nereids, sea spirits. Comparable categories of nature spirits are recognized in many traditional cultures throughout the world. What are we to make of them?

One suggestion, proposed by the archaeologist T. C. Lethbridge, is that they are not conscious entities as much as kinds of fields. The qualities and character associated with waterfalls, for example, he attributed to "naiad fields."[7] At first glance, this simply seems to involve a vague new terminology as obscure as the traditional one. But I think it is an idea worth pursuing. Fields are regions of influence, and in this general sense, the term is appropriate. But then what kinds of fields could the fields of places be? They are obviously not reducible to the known fields of conventional physics, though electromagnetic fields no doubt contribute something to the quality of the place. However, it might make sense to think of the fields of places as *morphic fields*. Such fields are associated with self-organizing systems at all levels of complexity, and they are ordered in nested hierarchies (Fig. 5.1). If particular places do indeed have morphic fields, then these fields must be embedded within larger fields, such as the fields of river systems and mountain chains, and these in turn within the fields of islands, archipelagoes and continents, and ultimately within the morphic fields of Gaia and the entire solar system.

When I first began to think along these lines, I was reluctant to extend the concept of fields to places because this seemed to be stretching the idea too far. But then I realized that the field concept itself is grounded in the idea of place. It involves a metaphorical extension of the everyday sense of fields as places of activity—as in cornfields, battlefields, football fields, and coalfields. The wider sense of fields as areas or spheres of action, operation, or investigation—as in the "field of trade" or "field of view"—predates by centuries the technical use of this term in physics. When the word was adopted by Faraday in the 1830s for his field theory of magnetism and electricity, he inevitably drew on these already-established usages, which go back to the Old English *feld* and *folde,* "earth" or "land."[8] Thus, a field theory of places recalls the fact that fields *are* places.

The idea of the spirits of places as morphic fields implies that particular places are subject to morphic resonance from similar places in the past. The generic qualities of places, traditionally expressed in terms of the various classes of nature spirits, will indeed have a kind of collective character and memory. Moreover, particular places will have their own memories by self-resonance with their own past. Morphic resonance takes place on the basis of similarity; hence the patterns of activity of the place in the summer will tend to resonate most specifically with those in previous summers, the winter patterns with previous winter patterns, and so on.

Memory also plays a part in the responses of animals and of people to the particular place. Obviously when people enter the place, their memory of their previous experience in the place or in similar places will tend to affect their present experience. But in addition to individual memory, through morphic resonance there will also be a component of collective memory through which a person can tune in to the past experiences of other people in the same place. Of course, not all such experiences are good. For instance, throughout the world it is widely believed that places where people have been murdered, executed, or tortured are inauspicious if not actually haunted.

Thus, in the context of morphic resonance, the experience of

THE REBIRTH OF NATURE

a particular place involves both a memory inherent in the place itself and a memory of previous experiences of the same individual and similar individuals in the place. The quality or atmosphere of the place does not depend just on what is happening there now but on what has happened there before and on the way it has been experienced. These principles are quite general, but take on a special significance in relation to places traditionally regarded as sacred.

Sacred Places

All over the world, certain places are regarded as sacred. Such holy places may be natural sites, such as springs, mountains, and groves of trees on hilltops, or they may be places where standing stones, stone circles, tombs, shrines, temples, churches, cathedrals, or other buildings have been erected. The choice of sites for these sacred structures traditionally depends on the particular qualities of the place. And their orientation often relates them to significant natural features, such as the point on the horizon where the sun rises at midsummer, or to other sacred places, as in the way that mosques are oriented toward Mecca. Sacred structures are often erected at places already hallowed; for example, many churches and cathedrals in Europe were built on pre-Christian sacred sites (Chapter 2).

Such sites are venerated because of what has happened there in the past. They are places where sacred experiences or revelations have occurred; where heroes and saints were born, lived, or died; or where their remains are preserved. What happened in the past can in some sense become present there again, and thus they can act as doorways to realms of experience that transcend the ordinary limitations of space and time. To traditional peoples,

> a sacred spot never presents itself to the mind in isolation.
> It is always part of a complexus of things which include

the plant and animal species which flourish there at various seasons, as well as the mythical heroes who lived, roamed or created something there and who are often embodied in the very soil, the ceremonies which take place there from time to time, and all the emotions aroused by the whole. (L. Levy-Bruhl, 1938)[9]

Holy places are generally separated from the profane world around them by a boundary. On crossing this boundary, people can share in the power of the place and hold communion with its sacredness; they can participate in the process by which it was consecrated in the first place.[10] Such beliefs are present in all religions, including Christianity, and sacred places are found all over the world, including North America:

Moses was instructed to take off his shoes so as to respect the holy ground on which he stood before the burning bush on Mount Horeb. Muhammad first heard the reciting voice of the angel Gabriel in the lonely cave on Mount Hira outside Mecca. Even amid the celebrated rootlessness of American culture, the phenomenon can be observed. The chapel at Valley Forge is a civil pantheon built on hallowed soil. The hill Culmorah in Palmyra, New York, is the sacred link of the Mormons to an ancient past. Native Americans of the Pacific Northwest still refer to Mount Rainer as Tahoma, "the Mountain that was God." Human beings are invariably driven to ground their religious experience in the palpable reality of space. (Belden Lane, 1988)[11]

The processes by which particular places become holy are still going on. A religious experience of the wilderness has endowed many of the American national parks with a transcendental quality. For many who visit them, they are more than recreational areas; they are natural temples or sanctuaries. And in Europe in modern times, visions of the Virgin Mary—for example, at Fatima in Portugal, at Lourdes in France, at Medjugorje in Yugoslavia—have resulted in these places becoming major centers of pilgrimage, famed for miraculous healings and continuing visionary experiences.[12]

Living in Harmony with Places

In societies in which the differing qualities of places are generally recognized, the sites of villages and towns, temples, houses, tombs, and other structures are not chosen just for convenience but for their harmonious relationship to their surroundings. The art of choosing such places is called *geomancy* (literally "earth divination"). The best-known contemporary form of geomancy is the Chinese system of *feng shui* ("wind and water"). Practitioners of this art command large fees in Hong Kong and other Chinese communities, and advise not only on the siting and orientation of buildings but also on the positioning of doors and windows and the layout and furnishing of rooms, so that the various movements and activities within them are in a harmonious relationship to one another and the surroundings.

Feng shui is based on the belief that in every place there are topographical features that indicate or modify the patterns of energy flow and that it is desirable to align human activities with these patterns. The forms of hills, the directions of watercourses, the prevailing winds, the heights and forms of surrounding walls and buildings, the positions of trees, roads, and bridges all play an important part, as does the influence of the sun, moon, planets, and stars.[13] Feng shui can be thought of as a system by which the fields of places are interpreted, enabling practical decisions to be made in the light of this understanding.

In Europe too, sites and orientations for ancient buildings were chosen on the basis of a combination of practical, intuitive, and symbolic considerations, although these were probably never elaborated as systematically as the principles of feng shui.

Modern researchers in "earth mysteries"—the hidden powers and qualities of places—have found that the location of major sacred sites is often associated with particular patterns of underground "energy flows" revealed by the techniques of dowsing, akin to water divining.[14] The nature of these "energy flows" is obscure, as is the basis of dowsing itself. But there is no doubt that some places are agreeable to live in, while other places create a sense of unease, and it would be foolish to rule out the possibility that underground patterns of activity could affect the quality of places.

In many parts of the world, once the site has been chosen for a new city or building, the first piercing of the ground is accompanied by carefully timed ceremonies intended to bring the new enterprise into harmony with the place and the cosmos. Even in England as late as 1675, the foundation stone of the Royal Observatory at Greenwich (from which the zero meridian of longitude is set) was laid at the exact moment elected as the most auspicious by Astronomer Royal John Flamsteed.[15]

The first piercing of the earth typically involved the driving of a stake into the ground, symbolizing the piercing of the head of the earth serpent, fixing it in place. This geomantic moment is one of the meanings implicit in the image of the hero, such as St. George, piercing the dragon. Traditionally, such events were accompanied by some form of sacrifice. It was commonly believed that the spirit of the place required the blood and body of a living creature as an offering for the sacrilegious act of breaking the virgin ground. Human or animal remains, sacred relics, or other objects were buried beneath or in the fabric, as offerings to the spirit of the earth. And the spirit of the sacrificed animal or person was believed to act as a guardian to the new building and ward off evil spirits. The relics of saints were buried in churches and cathedrals for these reasons, often beneath the high altar. Sacrificial remains of animals have been found beneath many old buildings in Europe, the most common being horses, oxen, and cats. As late as 1895 at Black Horse Drove in

Cambridgeshire, England, the head of a horse was interred in the foundations of a Methodist chapel, and in 1913 a cart horse was buried beneath the foundations of the terrace of the Arsenal football stadium in North London.[16]

In some parts of Europe, such as rural Switzerland, traditional builders' rites are still observed, and even today the laying of foundation stones of important buildings is accompanied by ceremonies. But for the most part, the modern world is suffering from geomantic amnesia.[17] The belief that the foundation of a new building should be accompanied by ceremonies to establish a harmonious relationship with the spirit of the place is generally dismissed as superstition. And when misfortunes befall those who have ignored these traditional principles, they are ascribed to medical and other scientifically recognized causes, or just bad luck.

The modern movement in architecture deliberately broke with the traditions of the past, approaching the construction of buildings in the spirit of mechanistic science. In Le Corbusier's memorable phrase, buildings were "machines for living in." Most scientifically trained town planners and architects know nothing of geomancy or traditional ideals of living in harmony with a place. And the results are only too apparent all around us.

From Tourism to Pilgrimage

Although the spirits of places elude analysis in terms of mechanistic science, their importance is implicitly recognized by tourists who visit famous sites because of their particular qualities and histories. Tourism is a vast modern industry, worth billions of dollars a year. Many of the places that act as magnets for tourists are ancient places of sacred power: in Britain, for example, Stonehenge, Westminster Abbey, Glastonbury, and Iona; in Egypt, the temples, tombs, and pyramids; in France, caves such as Lascaux and cathedrals such as Chartres; in Mexico, the temples of the Mayans; the living temples of India and Bali; the holy cities of Rome and Jerusalem; the sacred mountains of the Himalayas.

Tourism is like a secularized or unconscious form of pilgrimage. Indeed, many tourist attractions were places of pilgrimage in the past, and some still are. But whereas pilgrims visit a hallowed place as an act of religious devotion, tourists visit it as more or less detached spectators. Pilgrims participate in the sacred qualities of the place and the religious observances practiced there; tourists do not. Pilgrims add to the power of a sacred place; tourists subtract from it.

The primary factor in pilgrimage is intention. If we go as pilgrims to a sacred place, we go in the hope of being inspired or blessed, or to give thanks. We can inform our intention by learning the stories of the place and its spirit, and by hearing of other people's experiences there. The journey itself is as much a part of the pilgrimage as arriving, and by remembering that we are not going for the sake of comfort, we are better able to respond positively to any difficulties we may encounter.

The final approach is best made on foot—to experience a sense of the place and adjust to the ancient rhythm of walking. Often circumambulation of the sacred center is customary; walking around it is a recognition of its centrality. In most traditions, this is usually done in a sunwise or clockwise direction, but in some, such as the Bon of Tibet and Muslims at Mecca, counterclockwise. On entering the sacred center, one normally makes some offerings; for example, the lighting of candles or incense. Prayers may be offered. And something (such as holy water) may be brought back to be shared with those at home.

I believe that much good would flow from a change of attitude whereby tourists became pilgrims once again. Going to a sacred place as a tourist impoverishes the experience of the place, but going as a pilgrim enriches it. In our personal and collective lives, the transformation of tourism into pilgrimage has a large part to play in the resacralizing of the earth.

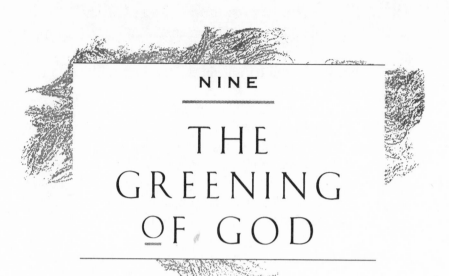

NINE

THE GREENING OF GOD

Once again it makes sense to think of nature as alive. The old cosmology of the world-machine, with the divine engineer as an optional extra, has now been superseded within science itself (Chapters 4–6). This completely alters the context in which the relationship between God and nature can be conceived. For if the entire cosmos is more like a developing organism than an eternal machine, then the God of the world-machine is simply out of date.

If nature is alive, she can be thought of as entirely autonomous, with no need for God. On the other hand, if God exists, he must be the God of a living world. In this chapter, I look at ways of thinking about nature, with and without God.

The Rediscovery of the God of the Living World

Over the last two decades, with the rise of the green movement and the growing awareness of the ecological crisis, members of different religious traditions have been engaged in a rediscovery of their spiritual relationship with the living world. One expression of this process was a pilgrimage to Assisi, the birthplace of St. Francis, organized by the World Wildlife Fund in September 1986. The pilgrims included Muslims, Christians, Jews, Buddhists, and Hindus, and the aim was "to celebrate the diversity of the world's response to nature through doctrine, teachings, symbol, art, drama, prayer, scriptures, music, tradition and mythology" (WWF pamphlet, 1986).

Within the Christian churches, the rediscovery of the God of the living world is currently taking place in several ways. One is through a revival of the animistic traditions that prevailed until the Protestant Reformation and the growth of the mechanistic theory of nature. The God portrayed in the Bible, in the teachings of Jesus, in the writings of the Church Fathers, and by medieval and Renaissance theologians was a God of living nature. The all-pervasive creative power of God worked not just in human lives but in the life of the earth and the cosmos. God was not separate from the course of nature and human history but immanent in them.[1] Here is how the twelfth-century abbess and mystic Hildegard of Bingen expressed this vision in one of her songs:

> I, the fiery life of divine wisdom,
> I ignite the beauty of the plains,
> I sparkle the waters,
> I burn in the sun, and the moon, and the stars,
> With wisdom I order all rightly. . . .
> I adorn all the Earth,
> I am the breeze that nurtures all things green . . .
> I am the rain coming from the dew
> That causes the grasses to laugh with the joy of life.
> I call forth tears, the aroma of holy work.
> I am the yearning for good.[2]

In addition to this rediscovery of the traditions of animistic Christianity, a related process of rediscovery is taking place through the experience of other religions. A number of Westerners have rejected the Christian religion and explored instead the religious traditions of the East, particularly Hinduism and Buddhism; others have followed the Sufi traditions that grew up within Islam; others have been drawn to the shamanic traditions of traditional peoples; others have attempted to revive aspects of pre-Christian paganism and the religion of the goddess. Such searches usually spring from a sense that Christianity and Judaism have lost contact with mystical insight, visionary experience, a sense of the life of nature, and the power of ritual.

I was so strongly drawn to India and its cultural and religious traditions that when the chance came to work there as a plant physiologist in an international agricultural research institute, I took it and spent about seven years in southern India. Much to my surprise, over a period of years, I found myself being drawn back to Christianity. Through being in India, I discovered the power of pilgrimage, ritual, seasonal festivals, meditation, and prayer. I saw these as living realities in the lives of my Hindu and Muslim friends and acquaintances, and in the lives of Indian Christians. They became living realities in my life too. I was much helped in the rediscovery of my own tradition by a Benedictine monk, Dom Bede Griffiths, who has lived for many years in a small ashram on the banks of the river Cauvery in Tamil Nadu.[3] I stayed in his ashram for a year and a half, and there wrote the first draft of my book *A New Science of Life*.

The Animistic Roots of Judaism and Christianity

In the late nineteenth and early twentieth centuries, it was fashionable to see the evolution of human consciousness in terms of an ascent from animism and a belief in magic, through a stage of religion with a belief in spirits and gods, to the advanced state of consciousness represented by science. Religion, though superior to primitive animism, was still pervaded by animistic and magical modes of thought. According to this view, science had superseded religion not only by virtue of its superior rationality but because it was a more effective way of manipulating the environment to serve human ends.

Anthropological scholars such as James Frazer showed with massive evidence that many aspects of Judaism and Christianity were strikingly similar to myths and beliefs found in many other religious and animistic traditions.[4] For example, the story of the virgin birth of Jesus has many mythological parallels, and the sacrifice, death, and resurrection of Jesus celebrated in the festival of Easter resembled the rites of gods such as Attis and Osiris, whose annual death and resurrection ensured the fertility of the earth. Yet more primitive were the practice of human sacrifice and the institutions of sacrificial kingship. Jesus was a sacrificial king, "The King of the Jews" written over his cross.

Frazer and other rationalists regarded this demonstration of the animistic, magical, and pagan roots of Judaism and Christianity as a way of dismissing these religions, rising above them through reason. And in my own scientific education, I found this one of the most persuasive arguments for seeing Christianity, like all religion, as essentially superstitious. Like Frazer, I thought that science represented a higher mode of consciousness. However, it now seems to me a strength of Christianity that it is grounded in an animistic experience of nature and incorporates archaic mythic themes. It includes and transforms archetypal patterns deep in our collective memory. Indeed, much of Christianity is incomprehensible without an appreciation of its mythological and "primitive" background, including the rites of animal and human sacrifice.

Shamanic Aspects of Judaism and Christianity

Like many other people, in the last two decades I have been fascinated by shamanic traditions, including those that involve the

use of psychedelic plants. *Shamanism* is the name given by anthropologists to practices of ecstatic visionary experience found
among traditional peoples all over the world. The roots of shamanism are archaic and are thought by some anthropologists to
go back almost as far as human consciousness itself. The mythologies of shamanic peoples, their symbolism and healing
techniques, are all based on the ecstatic experience.[5] Their common themes are "the descent to the Realm of Death, confron-
tations with demonic forces, dismemberment, trial by fire,
communion with the world of spirits and creatures, assimilation
of the elemental forces, ascension via the World Tree and/or the
Cosmic Bird, realization of a solar identity, and return to the
Middle World, the world of human affairs."[6]

Shamanism sheds much light on the Judeo-Christian tradition.
Above all, it illuminates the figure of Jesus himself, including his
death on the Tree, descent into hell, resurrection, and ascent into
heaven. It also enables us to see the archaic roots of visionary
revelation, inspired prophecy, initiation by baptism, and miraculous powers of healing.

In the Book of Samuel we read that "in Israel in days gone by,
when someone wished to consult God, he would say, 'Let us go
to the seer.' For what is nowadays called a prophet (*nabi*) used
to be called a seer" (1 Samuel 9:9). The institution of the seer of
the nomadic period of the Jews was modified after the conquest
of Palestine under the influence of the *nabiim,* the ecstatic
prophets of the Canaanite religion, such as the prophets of Baal
(1 Kings 18:19 ff; 2 Kings 10:19). The seers were not attached
to sanctuaries, whereas the prophets were.[7] For example, when
Samuel anoints Saul king of Israel, he gives him instructions for
the journey he is to take, including the following:

> Then when you reach the hill of God, where the Philistine
> governor resides, you will meet a company of prophets
> coming down from the shrine, led by lute, drum, fife and
> lyre, and filled with prophetic rapture. The spirit of the

Lord will suddenly take possession of you, and you too
will be rapt like a prophet and become another man.

(1 Samuel 10:5–6)

In the early church, the charismatic gifts of the Holy Spirit,
including the gifts of healing, speaking in tongues, and prophecy, were expressed in states resembling shamanic possession.
These gifts have been cultivated in the various Pentecostal sects,
and through the recent charismatic revival are now widely in-
voked within mainstream Christianity, including the Methodist,
Anglican, and Roman Catholic churches.

Visionary experience, sometimes induced by practices such as
fasting, is a recurrent feature of Christian mysticism and, like
the Hebrew prophetic tradition, has many precedents in the ecstatic visions of shamans. In this century, exotic psychedelic
forms of Christianity have grown up in America, where psy-
choactive plants traditionally used in indigenous shamanic
traditions are now taken ceremonially as a form of Christian
communion. One such church is the Native American church in
the Southwest, using the mescalin-containing cactus *peyote.* Another is currently spreading among the forest people of the Amazon, with a communion of *ayahuasca* or *daime,* a psychedelic
mixture of Amazonian plants.[8] The patroness of these Amazonian churches is the Virgin Mary in the form of the Queen of the
Forest.

Ritual initiations, such as the practice of baptism by total immersion in the Jordan by St. John the Baptist, were clearly effective in a way that was more than "just symbolic." Many of those
who were thus baptized had the experience of dying and being
born again, a phenomenon that is fundamental to initiation rituals all over the world. A similar process occurs spontaneously
in near-death experiences.[9]

These experiences, which are characterized by a common
pattern, include such elements as an overwhelming feeling

of peace and well-being, finding oneself out of one's body, floating or being propelled through a dark void, becoming aware of a brilliant white or golden light and encountering or communicating with a "presence" or "being of light," at which time one's fate is generally decided, seeing a panoramic review of one's life, entering a world of supernal beauty and recognizing deceased loved ones and conversing with them, and a number of other transcendental elements. The phenomenon usually has a profound effect on the person who has experienced it, not the least of which is a greatly diminished fear of death.[10]

It seems to me very probable that John the Baptist was a drowner.[11] If he held the initiates under water just long enough, they would indeed have had the life-changing experience of dying and being reborn again. Although in most Christian churches the practice of infant baptism through the sprinkling of water means that much of its original initiatory quality has been lost, the Baptists retain the practice of baptizing adults through the ritual of total immersion, and it is the Baptist churches that place the greatest emphasis on the experience of being born again. Indeed, their form of Christianity is centered on this conversion experience.

The Mother of God

The Virgin Mary is the predominant form in which the Mother has been honored and worshiped by Christians. She was proclaimed Mother of God at the Council of Ephesus in 431, and her cult spread rapidly throughout all Christendom. She progressively took on the titles and attributes of a variety of pre-Christian goddesses; many of her shrines were at places formerly consecrated to goddesses, as at Ephesus, the site of the great temple of Artemis. This process took place all over Europe and continued in Latin America and other areas evangelized by

Roman Catholics. In Mexico, for example, only ten years after the Spanish conquest, she appeared in a vision to an Aztec convert in the form of the Virgin Mary of Guadalupe and asked for a church to be built at the exact place on which had stood a temple of Tonantzin, the Aztec mother goddess.[12]

Our Lady of Guadalupe is a black Madonna, both in her Mexican and original Spanish forms. Black or dark forms of Mary have been the object of deep veneration in many parts of Christendom, including Walsingham in England and Chartres in France. Such dark forms of Mary are believed to have many miraculous powers, including the gift of fertility to barren women. The original Virgin of Guadalupe, in Spain, was herself a miracle worker. Her ancient image is said to have been found in a cave by a shepherd, to whom she had appeared by the river Guadalupe in the early fourteenth century. Her shrine was established at this very cave, where it remains to this day. Both Columbus and Cortés made pilgrimages to it before setting sail for America.[13]

No one knows how the tradition of black Virgins first arose, but their symbolic significance must depend in part on their association with the earth and with death.[14] The Great Mother of archaic religions was the source of life and fertility, and the womb to which all life returned. The Indian black goddess, Kali, is the Great Mother in her aspect as destroyer, but also as the source of new life. And the Virgin Mary is associated with death as well as fertility and nourishment: "Holy Mary, Mother of God, pray for us sinners now and at the hour of our death."

The cult of Mary remains an essential feature of Orthodox and Catholic Christianity. But it was suppressed in Protestant countries during the Reformation; her shrines were desecrated, and devotion to her was denounced as a form of idolatry (Chapter 1). Although the more extreme Protestant sects maintain their opposition to "mariolatry," Marian shrines, pilgrimages, and devotions have gradually been restored in the Anglican church.

The cult of Mary has recently been much revived and strengthened by Pope John Paul II, himself an ardent devotee of the miracle-working Black Madonna of Czestochowa, who was consecrated as queen and patroness of Poland by King Jan Casimir in 1656. After becoming pope, he made a pilgrimage in 1979 to her shrine, at which he consecrated her Mother of the Church "while the second millennium of the history of Christianity on Earth is about to draw to a close." He concluded his prayer to her as follows:

> How many problems, Mother, should I not present to you by name in this meeting. *I entrust them all to you,* because you know them best and understand them.
>
> I entrust them to you in the place of the great consecration, from which one has a view not only of Poland but of the whole church in the dimensions of countries and continents—the whole church in your maternal heart.
>
> I who am the first servant of the church offer the whole church to you and entrust it to you here with immense confidence, Mother. Amen.[15]

Throughout the 1980s, visions of the Virgin Mary have been seen persistently in several parts of the world, most notably at Medjugorje in Yugoslavia where they began in 1981 on St. John the Baptist's Day, June 24. There she announced herself as Queen of Peace. Many of her messages are about the world crisis; she calls on people to believe in the grace of God before darkness overcomes the world.[16] The pope proclaimed a special Marian year 1987–1988, at which special devotions were observed throughout the Catholic church, and many prayers addressed to her.

The pope's honoring of the earth as mother is expressed through his custom of kissing the ground as soon as he alights from planes. It seems a shame that this is usually tarmac.

Nature without God

In a mechanistic world, nature worship makes no sense. There is no point in trying to form a personal relationship with blind mechanical processes or with blind chance. All that matters is to try to understand nature so that it can be controlled for human ends. By contrast, in a living world, nature contains living powers far greater than human powers. In the cosmic evolutionary process and in the evolution of life on earth, she is vastly more creative than man. She is the source of life, and she brings forth its myriad forms with inexhaustible creativity. She is all material processes; she is the cosmic flow of energy; she is in all physical fields; she is chance and merciless necessity. Indeed, if there is no God, she is everything.

The trouble is that if nature is the Great Mother and there is no Father, the feminine principle is entirely predominant. An image of the Great Mother as all-powerful is just as unbalanced as an all-powerful Great Father. Some radical feminists and some male chauvinists may like the idea of the cosmic primacy of their own gender, but the metaphors of motherhood and fatherhood inevitably work against a one-sided view. When some people claim that everything comes from the Mother and others claim that everything comes from the Father, there is another obvious possibility: everything comes from both. This is in fact the traditional view in most parts of the world. If the earth is the realm of the Mother, the heavens are the realm of the Father, and all life depends on their relationship. Or if the feminine principle is the cosmic flux of power and energy, the masculine is the source of form and order, like Shakti and Shiva in Indian Tantrism. Or, as in the Taoist view, there is a continuous interplay of the feminine and masculine principles, yin and yang, throughout all nature. And there are many other ways of modeling polarities in terms of female and male, mother and father.

The image of nature as the Great Mother has been associated for millennia with sky-gods. If this male principle is suppressed, if God is eliminated from the picture, he is likely to live on in an unconscious form. This is the case in the mechanistic worldview, where nature is governed by eternal laws that transcend the physical world, laws that are the ghost of the rational, mathematical God of the world-machine. If nature is to be conceived of without God, and without God-substitutes in the form of disembodied laws, then nature must include both male and female principles within herself—or, rather, *itself*. For if nature is all and everything, it cannot be just female or just male but must include and embrace all polarities.

Everywhere we look in the realm of nature we find polarities, such as electrical and magnetic polarities. These can, if we like, be modeled in terms of gender; for example, positive electrical charge is associated with dense, relatively immobile atomic nuclei, a bit like eggs; negative charge is associated with the smaller electrons, moving in swarms, a bit like sperm. But sexual gender is only one of many kinds of natural polarity and only one of the ways we experience polarity in our own lives. Others include the polarities of up and down, in and out, front and back, right and left, past and future, sleeping and waking, friend and foe, sweet and sour, hot and cold, pleasure and pain, good and bad.

On the cosmological level, the primary polarity is between the expansive impulse that underlies the growth of the universe and the contractive field of gravitation that holds everything together. If the centrifugal force is predominant, the universe will expand indefinitely; if the centripetal, the universe will sooner or later stop growing and begin to contract until everything is annihilated in the Big Crunch. No one knows what will happen. But in the meantime, the interplay between these expansive and contractive principles underlies the processes of cosmic evolution.

As the newborn universe expanded and cooled, the primal unified field gave rise to the fundamental fields of gravitation, the quantum fields of material particles, and the electromagnetic field (p. 131). With further expansion and cooling, galaxies and stars came into being under the influence of gravitation, and within the stars the evolution of the chemical elements contin-ued. Later still, when matter ejected from exploding stars aggre-gated gravitationally into planets, a great variety of molecular and crystalline forms arose, and liquids, such as water, ap-peared for the first time. Then life emerged, at least on earth, and biological evolution began. The creative processes of evo-lution continue to this day and are expressed in our own collec-tive and personal lives. Creativity was inherent in the universe from the beginning. What is the nature of this evolutionary cre-ativity?

Evolutionary Creativity

The cosmic evolutionary process has a direction, an arrow of time. This arrow ultimately depends on the expansive impulse inherent in the cosmos since its birth. But because the growth of the universe has been accompanied by the development of fields, particles, atoms, galaxies, stars, planets, molecules, crystals, and biological life, the arrow of time has a cumulative developmental quality as well. Just as an embryo passes through a series of stages, each of which forms the foundation for the next, so does the evolutionary cosmos. There could be no biological life until there were planets, no planets until there were galaxies and stars, no galaxies and stars until there were atoms of matter, and no atoms of matter until their constituent particles had first come into being.

According to the hypothesis of formative causation, each new pattern of organization (a molecule, a galaxy, a crystal, a fern, or an instinct) involves the appearance of a new kind of morphic field. Through repetition, these new patterns of organization become increasingly habitual. Because of this habit-memory in-herent in nature, the evolutionary process is cumulative; new patterns of organization come into being in the context of the existing habits of nature and through repetition in turn become habitual. But if evolutionary creativity—the appearance of new patterns of organization—involves the coming into being of new kinds of morphic fields, where do these fields come from? Here we come back to the mystery of creativity, and as so often happens, there are three kinds of theory.

According to the first theory, all creativity emerges from the mother principle: it is inherent in nature, and emerges from blind, unconscious processes such as the workings of chance. It wells up from material activities. New patterns of organization, new morphic fields, spring into being spontaneously.

The second theory proposes that all creativity comes from the father principle. It descends into the physical world of space and time from a "higher," transcendent level that is mindlike. In the Platonic tradition, this eternal intelligence is the source and abode of the ideas reflected in the world of nature; for Christian Platonists it is none other than the mind of God. For Pythagoreans, this transcendent mindlike realm is mathematical. From mathematical principles, the universe comes into being and evolves; by mathematical principles, everything is governed. Insofar as new kinds of fields come into being, they are governed by field equations that exist eternally in the transcendent mathematical reality, irrespective of whether they are reified in the physical world or not. Evolutionary creativity involves the manifestation in physical form of mathematical structures that have always existed, or rather that are beyond time altogether.

Third, there is the theory that all creativity comes from an interplay between the mother and the father principles, or, more abstractly, from below and above. It depends on chance, conflict, and necessity, the mother of invention. It arises in particular environments, at particular places and times; it is rooted in the ongoing processes of nature. But at the same time it occurs within the framework of higher systems of order. For example, new species arise within ecosystems; new ecosystems within Gaia; Gaia within the solar system; the solar system within the galaxy; the galaxy within the growing cosmos. At every level of organization, there is a higher level that includes it, right up to

THE GREENING OF GOD

the level of the cosmos. Many theoretical physicists now think that the fundamental fields of physics arose from a higher and more inclusive field, the primal unified field of the universe; in the same way, at every level of organization, new morphic fields may arise within and from higher-level fields. Creativity occurs not just upward from the bottom, with new forms arising from less complex systems by spontaneous jumps; it also proceeds downward from the top, through the creative activity of higher-level fields.

The same principles apply to human creativity. It depends on accidents, conflicts, and needs, and is rooted in particular bodily, psychological, cultural, and environmental processes. At the same time, new inventions, new insights, new works of art, come into being in the context of ecologies, societies, cultures, and religions, and ultimately in the context of Gaia, the solar system, the galaxy, and the cosmos—and, as many creative people have themselves thought, God. Traditional theories of human creativity ascribe it to inspiration from a higher source working through the creative individual, who acts as a channel. The same conception underlies the notion of genius; originally the genius was not the person himself but his presiding god or spirit.

As with all polarities and dualities, when we try to conceive of how they are unified or held together, sooner or later we arrive at the idea that they are aspects of a higher unity. And indeed, according to the Big Bang theory, fields and energy arose together within the original cosmic singularity. All physical phenomena—such as sunlight, molecules, trees, and stars--have both a field aspect and an energy aspect. The wave and particle aspects of light, for example, are not two separate things but two aspects of the same structure of activity. So are the field and energy aspects of everything else, including ourselves.

A view of nature without God must account not only for the many kinds of polarity inherent in the world and our experience but also for the higher form of unity that includes these poles. Indeed, since the universe is a unity by definition, there must be a unitary principle that includes all nature. What is the nature of this unity? It is not enough to conceive of it as static, for the universe is evolving. Somehow the underlying unity of nature has to give rise spontaneously to new kinds of organisms and new patterns of behavior, which are themselves unities, wholes, or holons.

Thus a view of nature without God must include a creative unitary principle that includes the entire cosmos and unites the polarities and dualities found throughout the natural realm. But this is not far removed from views of nature *with* God.

Creative Trinities

An understanding of evolutionary creativity in terms of the interaction of two principles, such as fields and energy, inevitably implies a third, unifying principle of which they are both aspects. This unity is implied in the sexual metaphor; the generative power of father and mother depends on their union, and their offspring unite aspects of both parents. This is expressed in the most direct way in Indian tantric images of Shakti and Shiva in sexual embrace; in a more abstract manner in the Taoist representation of yin and yang intertwined and interpenetrating within a circle unifying both, the Tao (Fig. 9.1). In other trinities, the polarity of gender is replaced by different principles, as in trinities of goddesses and gods such as the Hindu trinity of Brahma the creator, Vishnu the preserver, and Shiva the destroyer. Here Vishnu could represent the organizing fields of nature, Shiva the cosmic flux of energy, and Brahma the creative unity that includes both.

The Christian conception of God is as a creative trinity: the Father, Son, and Holy Spirit. The mystery of the Holy Trinity has traditionally been considered in a variety of ways. One is the psychological model, favored by St. Augustine.[17] Here the Father is the knower, the Son the known, and the Spirit the relationship between them, the bliss of knowing. Another model is implicit

Figure 9.1. The interplay of yin and yang, the feminine and masculine principles, enclosed within the unity of the Tao.

in the identification of the Son with the Word, or Logos. Clearly this biblical concept of the Word relates to the spoken rather than the written word and thus implies not only a vibratory pattern of physical activity but also an unfolding pattern of meaning. And just as human speech involves imposing an ordered pattern of vibrations on the outward flow of the breath, so the creative Word of God works together with the breath of God, the onward and outward movement of the Spirit. The Spirit is the principle of flow and change. The traditional images of the Spirit include breath, wind, life, flames, and the flying dove. Its movement is free and unpredictable: "The wind blows where it wills, you hear the sound of it; but you do not know where it comes from or where it is going" (John 3:8).

In the doctrine of the Holy Trinity in the Orthodox churches, the Spirit plays a richer and fuller role than it does in most Western theology and gives a much stronger sense of the immanence of nature in the divine:

The creative energies of God did not merely produce the created world from without like a builder or engineer, but

are the ever-present, indwelling and spontaneous causes
of every manifestation of life within it, whatever form this
may take. [This understanding] depends, in other words,
upon the recognition of the continuing, vitalizing activity
of the Holy Spirit in the world, animating these energies—
luminous uncreated radiations of the divine—in the very
heart of every existing thing. (Philip Sherrard, 1987)[18]

In the context of evolutionary cosmology, the Spirit underlies
the onward flow of energy and the expansive impulse of the
universe; the Word is in the patterns of activity and meaning
expressed through fields. God the Father is the speaker, the conscious source of both Word and Spirit who transcends both.
Thus, the energy and fields of the evolutionary cosmos have a
common source, a unity. And not just a unity but a conscious
unity.

If the fields and energy of nature are aspects of the Word and
Spirit of God, then God must have an evolutionary aspect, evolving along with the cosmos, with biological life and humanity.
God is not remote and separate from nature, but immanent in
it. Yet at the same time, God is the unity that transcends it. In
other words, God is not just immanent in nature, as in pantheist
philosophies, and not just transcendent, as in deist philosophies,
but both immanent and transcendent, a philosophy known as
panentheism. As the fifteenth-century mystic Nicholas of Cusa
put it: "Divinity is the enfolding and unfolding of everything
that is. Divinity is in all things in such a way that all things are
in divinity."[19]

The creative polarity of Spirit and Word, like other creative
polarities, can be modeled in terms of gender, but in an ambig-
uous manner. If we think of the feminine principle as active, like
Shakti, then the Spirit is feminine, the Word masculine. The
word for "spirit" in Hebrew, *ruah,* is indeed feminine.[20] (In
Greek, the corresponding word, *pneuma,* is neuter; in Latin it is
masculine, *spiritus.*) Alternatively, taking the masculine princi-

ple as active, then the Spirit is masculine and the Word feminine.
This is an unfamiliar way of thinking, given the identification of
the Word with the Son. But there is no doubt that the biblical
concept of the Word of God has much in common with the feminine divine Wisdom, Sophia.[21] In the Book of Proverbs, she
speaks of herself as follows:

> The Lord created me in the beginning of his works
> before all else that he made, long ago.
> Alone, I was fashioned in times long past,
> at the beginning, long before earth itself. . . .
> When he set the heavens in their place I was there . . .
> when he prescribed the limits for the sea
> and knit together earth's foundations.
> Then I was at his side each day,
> his darling and delight,
> playing in his presence continually,
> playing on the earth, when he had finished it,
> while my delight was in mankind.
>
> (Proverbs 8:22–3, 27, 29–31)

In the prologue to St. John's Gospel, the Word is very like
Wisdom:

> In the beginning the Word already was. The Word was in
> God's presence, and what God was, the Word was. He
> was with God at the beginning, and through him all
> things came to be; without him no created thing came
> into being. In him was life, and that life was the light of
> mankind.
>
> (John 1:1–5)

The Gender of God

When God is conceived of in relation to Mother Earth or
Mother Nature, he is male. The Father and the Mother create

together. In the first chapter of the Book of Genesis, the primal mother principle is the void or the deep, coexistent with God from the start. From this womb everything ultimately comes forth, through a series of divisions made by God: the light from the darkness, the day from the night, the heaven from the earth, the seas from the dry land. The power to give birth is implicit in the primal void or deep, and present in the earth and the seas. When God calls forth plants and animals, he does not create them directly; they are born from the earth and the waters, from the womb of the Mother (p. 19).

But when God is conceived of as creating everything from nothing, as in the theology of St. Thomas Aquinas, the mother principle has to come forth from God or be inherent within God. In this sense, God is both Mother and Father, male and female. We have already seen how this male-female polarity can be expressed in terms of the Word and the Spirit within the Holy Trinity. But then is God the Father also God the Mother? Is the Godhead, the source of the Holy Trinity itself, masculine, feminine, or neuter?

The earliest Hebrew name for God was *elohim,* a plural word of obscure origins that could mean "goddesses" and "gods," and was also used to mean the "spirits of the ancestors."[22] Nevertheless, by convention, God is treated as masculine, and so it is both surprising and illuminating to encounter the feminine imagery employed by medieval mystics such as Meister Eckhart. "God lies in the maternity bed, like a woman who has given birth, in every good soul which has abandoned its self-centeredness and received the indwelling of God."[23] The hermitess Julian of Norwich wrote of the maternal aspect of the Godhead enclosing us in "the deep Wisdom of the Trinity [who] is our Mother."[24]

Just as an all-powerful nature cannot be simply female, neither can an all-powerful God be simply male. Either we have to think of the male-female polarity as coexistent from the start, or we have to derive them both from a common source that transcends the polarity between them. Even if we want to try to conceive of the universe in purely scientific terms, the same questions arise. Science, like religion, is pervaded by a strong sense of a fundamental unity. This intuition underlay Einstein's search for a unified field theory and currently inspires attempts to conceive of the primal field of the cosmos and the primal source of energy. Here science meets theology; for if fields and energy have a common source that transcends both, we find ourselves back in the field of creative trinities. And as theology meets science, a new evolutionary conception of the creative trinity is coming into being; theology itself is evolving.

An Evolutionary God

In a Christian context, fields can be thought of as an aspect of the Word and energy as an aspect of the Spirit. If the Word and Spirit of God are immanent in the realm of nature and immanent in the creative process, then God must be evolving along with nature. At the same time, God somehow gives this process its overall purpose, which the evolutionary mystic Teilhard de Chardin conceived of as the Omega Point, the state of unity toward which everything is developing. Such a conception is necessarily obscure since it goes beyond anything that has happened so far, beyond our limited powers of thought. This is how Teilhard described it:

> By its structure Omega, in its ultimate principle, can only be a distinct Center radiating at the core of a system of centers; a grouping in which the personalization of the All and personalizations of the elements reach their maximum, simultaneously and without merging, under the influence of a supremely autonomous focus of union.[25]

New forms of theology have recently been developing in an attempt to conceive of the God of a living, evolutionary cosmos.

Evolutionary theology involves a radical break with traditional theological ideas of God as timeless, uninfluenced by events in the world, acting on it but not really interacting with it. However, the God of the Bible was intimately involved with the history of the world and humanity. This remote, impassive image is not biblical but developed in the early church under the influence of Greek philosophy. In the spirit of Platonism, the mind of God was identified with the transcendent realm of eternal Forms; under the influence of Aristotle, God was conceived of as the unmoved mover.

By contrast, in the new, evolutionary view of God,

> like all living things, God not only acts on others, but also takes account of others in the divine self-constitution. . . . God is not the world, and the world is not God. But God includes the world, and the world includes God. God perfects the world and the world perfects God. There is no world apart from God, and there is no God apart from some world. Of course there are differences. Whereas no world can exist without God, God can exist without *this* world. Not only our planet but the whole universe may disappear and be superseded by something else, and God will continue. But since God, like all living things, only perfectly, embodies the principle of internal relations, God's life depends on there being some world to include. .
> (C. Birch and J. B. Cobb, 1981)[26]

Mystery

Each of us, faced with the mystery of our existence and experience, has to try to find some way of making sense of it. We have a choice of philosophies: the mechanistic theory of nature and of human life, with God as an optional extra; the theory of nature as alive but without God; or the theory of a living God together with living nature. Each of these views can be elaborated intellectually, each can be defended on rational grounds, and each is held with deep conviction by many people. In the end, we have to choose between them on the basis of intuition. Our choice is influenced by our acknowledgment of mystery and in turn affects our tolerance of it. Those with the lowest mystery-tolerance thresholds are drawn to the mechanistic-atheistic worldview, which as a matter of principle denies the existence of mysterious entities like souls and God, and portrays a disenchanted, unmagical reality proceeding entirely mechanically. Those who acknowledge the life of evolutionary nature admit the mystery of life and creativity. And those who acknowledge the life of God are consciously open to the mystery of divine consciousness, grace, and love.

TEN

LIFE IN
A LIVING
WORLD

From Humanism to Animism

What difference does it make if we think of nature as alive rather than inanimate? First, it undermines the humanist assumptions on which modern civilization is based. Second, it gives us a new sense of our relationship to the natural world and a new view of human nature. Third, it makes possible a resacralization of nature.

The humanistic vision has been a source of hope and inspiration in capitalist and communist countries alike, the dream of worldwide human progress and material development, of humanity living in peace and prosperity in a technological wonderland. In this humanistic heaven on earth, religious sanctions governing human behavior will be replaced by a rational humanitarian code of ethics, and humanity will consciously take the helm of the evolutionary process, engineering it for the greatest human benefit. In the late nineteenth century, for example, T. H. Huxley summarized his vision of human social progress as "a checking of the cosmic process at every step and the substitution for it of another, which may be called the ethical process."[1] Sigmund Freud took a darker view:

> Against the dreaded external world one can only defend oneself by some kind of turning away from it, if one intends to solve the task by oneself. There is, indeed, another and better path: that of becoming a member of the human community, and, with the help of a technique guided by science, going over to the attack against nature and subjecting her to the human will.[2]

From this humanist point of view, we are essentially alien to the larger living community; we need to subject it to ourselves lest we are subjected to it. In man, the natural world goes beyond itself into a new and more sublime form of grandeur. But neither Huxley, Freud, nor the other apostles of humanism had any idea of the disastrous consequences of such an attitude on the integral functioning of the earth, or on human destiny.

> The consequences are now becoming manifest. The day of reckoning has come. In this disintegrating phase of our industrial civilization, we now see ourselves not as the splendor of creation, but as the most pernicious mode of earthly being. We are the termination, not the fulfillment of the earth process. If there were a parliament of creatures, its first decision might well be to vote the humans out of the community, too deadly a presence to tolerate any further. We are the affliction of the world, its demonic presence. We are the violation of the earth's most sacred aspects. (Thomas Berry, 1988)[3]

The old dream of progressive humanism is fading fast. There are still those who dream of the conquest of the biosphere by the technosphere, the human control of biological evolution through genetic engineering, and so on. But attitudes are changing around, and within, many of us: there is a shift from humanism to animism, from an intensely man-centered view to a view of a living world. We are not somehow superior to Gaia; we live within her and depend on her life.

The green movement is much influenced by the Gaian perspective while trying to maintain a combination of humanistic and animistic attitudes. For example:

> The Earth is all we have—a world of finite resources. We depend on our planet and those resources for our very survival. We are part of a fragile interdependent network of Life. If our planet dies, we die. . . . The way we live now cannot go on forever. We must change or face extinction. (British Green Party pamphlet, 1989)

Conflicts inevitably break out when choices have to be made between people and the interests of endangered species, fragile ecosystems, or unspoiled wilderness. Humanistic greens put human interests first while trying to minimize the damage to the environment. Indeed, such an attitude is fast becoming acceptable in orthodox political circles. By their very nature, politicians have to compromise.[4] And some are already trying to establish a new consensus in which economic development remains the overall goal but is henceforth to be achieved by "sustainable" means that meet "the needs of the present without compromising the ability of future generations to meet their own needs."[5]

On the other hand, animists put the interests of Gaia first. Some are even prepared to contemplate the inevitability of huge reductions in the human population through war, pestilence, famine, floods, and other disasters. Such thoughts are deeply repugnant to humanist sentiments and can easily evoke accusations of misanthropy, even fascism.

This same humanist-animist debate that is going on so vigorously in the context of green politics has been raging in a more abstract form among ecological theoreticians. Some are primarily concerned with the need for changes in the human social order, seeing the present ecological crisis as a result of militarism, patriarchy, racism, and other forms of social domination.[6] But from the animistic point of view, social ecology is still too human-centered, still too shallow. Instead, "deep ecologists" advocate an ecology that is life-centered rather than human-centered, biocentric rather than anthropocentric, an ecology that recognizes the interconnectedness of all life and sees humanity as part of a larger living whole.[7]

But social ecology and deep ecology are not mutually exclusive. Social change and a change in our collective relationship to the earth will have to go together. Indeed, the needs of humanity include far more than material needs and can only be met if we live in an appropriate relationship to the living world around us.

The recognition that we need to change the way we live is now very common. It is like waking up from a dream. It brings with it a spirit of repentance, seeing in a new way, a change of heart. This conversion is intensified by the sense that the end of an age is at hand.

The New Millennium

At present there is a widespread and all-pervasive sense of crisis—environmentally, politically, economically, and socially—a belief that we have reached a turning point for our civilization, our species, and all life on Earth. We are at the same time approaching a major historical turning point, a new millennium in the Christian era. Given the millennial aspect of Christianity itself, this is in any case bound to create the expectation that we are nearing the end of an age. In the last book of the Bible, the Revelation of St. John the Divine, the ending of the age of history (the Apocalypse) occurs amid catastrophes, woes, and plagues, most of which seem only too familiar today, with no need for angels to administer them.

> I heard a loud voice from the sanctuary say to the seven angels, "Go and pour out the seven bowls of God's wrath on the earth."
> The first angel went and poured his bowl on the earth; and foul malignant sores appeared on the men that wore the mark of the beast and worshiped its image.
> The second angel poured out his bowl on the sea; and the sea turned to blood like the blood from a dead body, and every living thing in it died.
> The third angel poured out his bowl on the rivers and springs, and they turned to blood. . . .
> The fourth angel poured out his bowl upon the sun; and it was allowed to burn people with its flames. . . .
> The fifth angel poured out his bowl on the throne of the beast; and its kingdom was plunged in darkness. . . .
> The sixth angel poured out his bowl on the Great River, the Euphrates; and its water was dried up to prepare a way for the kings from the east.
> The seventh angel poured out his bowl on the air; and out of the sanctuary came a loud voice from the throne, which said, "It is over!" There followed flashes of lightning and peals of thunder, and a violent earthquake, so violent that nothing like it had ever happened in human history.
> The great city was split in three, and the cities of the nations collapsed in ruins.
>
> (Revelation 16:1–19)

One particularly mysterious prophecy has been puzzled over for centuries:

> The third angel blew his trumpet. A great star shot down from the sky, flaming like a torch, and fell on a third of the rivers and springs; the name of the star was Worm-

wood. A third of the water turned to wormwood, and great numbers of people died from drinking the water because it had been made bitter.

(Revelation 8:10–11)

In the Soviet Union, a great deal of interest was aroused by the fact that the Ukrainian word for "wormwood" is *chernobyl*. Meanwhile, in the United States, the great war in heaven between the dragon and Michael and his angels (Revelation 12:7) found a half-conscious echo in Ronald Reagan's dream of Star Wars.

Throughout the ages, people have looked for parallels in the signs of their times to those marking the end of the age, and they have often found them. But today the signs are striking enough to attract almost everyone's attention, with no need for visionary revelations, or biblical exegesis. There are many disaster scenarios on the market. Anyone can make up his own, putting together various combinations of factors like the population explosion, ecological devastation, pollution, the nuclear threat, droughts and climatic changes, new diseases, drug addiction, social disintegration, economic collapse, war.

Faced with this sense of impending doom, we need a spirit of repentance that is not just individual but collective. The imbalances that threaten the world are not the fault of a few greedy people in power; we are all part of the economic and political systems that have proved so destructive. At the very least, our attitudes and political and economic systems will have to change radically if we are to live in greater harmony with Gaia. The only question is how radically?

Some people hope that disaster can be averted through moderate reforms like the introduction of unleaded gasoline and catalytic converters in cars, paying more attention to the environmental impact of development projects, more recycling, a gradual shift toward renewable sources of energy, stricter controls on pollution, and a tax on energy consumption. Others place their faith in the power of green consumerism to influence the economy through market forces. Others think in terms of scientific "planet management."[8] At the other extreme are those who liken such reforms to rearranging the furniture on the *Titanic*. In their view, the present political and economic order is doomed. Just how it will break down and what will take its place is a matter of speculation, but there is not much that we can do about it unless we retreat to remote places and start training as survivalists. Most pessimistic of all are those who think nature is now so sullied by human activity that we can no longer form a relationship with her, even if we want to:

> The end of nature probably also makes us reluctant to attach ourselves to its remnants, for the same reason that we usually don't choose friends from among the terminally ill. . . . I find now that I like the woods best in winter, when it is harder to tell what might be dying. . . . I love winter best now, but I try not to love it too much, for fear of the January perhaps not so distant when the snow will fall as warm rain. There is no future in loving nature. (Bill McKibben, 1990)[9]

Such gloom is disempowering. The earth is not terminally ill, although our civilization may be. The evolution of life has survived catastrophes before, such as that which killed off the dinosaurs and countless other species, and will doubtless survive even if humanity does not. The cause of the present sickness is our modern technological civilization and its underlying ideologies. If we are to enter the new millennium with any hope for the future, we need to recover a new vision of human nature and of our relationship to the living earth.

Remembering Our Connections with Nature

We are continually reminded of nature's terrifying and destructive aspects through disasters such as earthquakes, hurricanes, floods, and droughts. Through the media, we are brought daily news and graphic details of other people's misfortunes; natural disasters are supremely newsworthy. Closer to home, the aspect of nature that concerns us most is the weather. Here, even in official reports, it is hard to avoid the sense that the atmosphere has a life of its own. For example:

> In the southern USA frontal systems were very active as arctic air sweeping south from Canada and warm humid southwesterlies from the Gulf of Mexico battled for supremacy. Heavy rain and thunderstorms in southeast Louisiana produced widespread flooding. . . . On Wednesday and Thursday tornadoes were reported in Mississippi and Alabama, together with hailstones nearly 8 cm in diameter.[10]

For those of us who live in cities, it is easy to forget the natural sources of our sustenance: it comes to us from shops or through wires and tubes. It is easy to forget where our wastes go: they disappear down drains, or garbage collectors take them away. The green movement has helped to remind us of the sources of our food, water, energy, and raw materials, and the destruction our demands engender. We are likewise becoming more aware of the sheer quantity of refuse we generate and the pollution of the air, waters, and land we are causing. It is no longer so easy to forget that we live on a finite planet with finite resources or to ignore the fact that Gaia is influenced by our activities and responds to them.

Evolutionary biology reminds us of our affinities with primates and other animals, and ultimately of our kinship with all life on Earth. Human consciousness must have developed with an awareness of the habits of animals that we hunted, the qualities of plants that we gathered, the seasonal changes of nature, and the characters of domesticated animals such as dogs. With the large-scale domestication of plants and animals initiated in the Neolithic revolution, a new familiarity was established with plants like barley, wheat, beans, hemp, and grapes, and with animals such as sheep, pigs, cattle, camels, and horses.

Our close association with domesticated animals and plants continues to this day. Those who breed, train, ride, and use horses, for example, come to know them intimately and often develop an intuitive communication with them. Anglers acquire a wealth of experience of the habits of fish, gamekeepers of game birds, trainers of performing animals, and so on. Many of those who breed and grow crops, vegetables, fruit trees, and garden plants come to recognize their habitual patterns of growth and their characteristic responses to the weather, the soil, and diseases and pests. Many people form relationships with plants, and some even talk to them.

Even modern city dwellers still have a deep need for personal connections with plants and animals. In Britain, for example, millions of people keep dogs, cats, or other pets; there are hundreds of thousands of pigeon enthusiasts, who often develop a close relationship with the birds they breed and race. Many millions of households have gardens, often lovingly tended, and still more homes have pot plants.

In Charles Darwin's time, there was no great division between serious scientific inquiry and natural history, largely the province of amateurs. Darwin himself was a natural historian and lived as a private gentleman with no academic post. However, the professionalization of biology that began in the late nineteenth century has now created a wide gulf between academic scientists and natural historians, studying various aspects of the natural world for the love of doing so. The knowledge and understanding of the naturalist is generally considered to be inferior to that of the professional scientist. But it seems to me that the opposite is true; the knowledge of the naturalist, which comes from an intimate relationship with nature, is deeper and truer than the kind obtained by detached mechanistic analysis. Of course, ideally, the direct experience of the naturalist and the systematic investigations of the professional scientist can complement and illuminate each other. Contemporary examples of such a synthesis are the study of bird migration, which involves a fruitful collaboration between professional scientists and amateur ornithologists, and the wonderfully illuminating accounts of the development of the English countryside by the botanist Oliver Rackham.[11]

Knowledge gained through experience of plants and animals is not an inferior substitute for proper scientific knowledge: it is the real thing. Direct experience is the only way to build up an understanding that is not only intellectual but intuitive and practical, involving the senses and the heart as well as the rational mind. Scientific investigations can illuminate and enrich this direct practical knowledge, but they are no substitute for it; indeed, they depend on it themselves.

The Reality of Mystical Experience

It is important to recognize the reality of our own direct experiences of nature in the wilderness, in the countryside, in forests, on mountains, by the sea, or wherever we have felt ourselves to be in connection with the greater living world. In its stronger forms, this sense of communion has the force of mystical experience, of illumination, surprise, and joy. But when we return to our everyday lives, we have a strong temptation to dismiss such experience as merely subjective, something that just happened inside us and did not involve any real participation in a life greater than our own. I think we should resist this temptation. Our direct intuitive experiences of nature are more real and more direct than mere theories, which go in and out of fashion. T. H. Huxley said much the same thing when discussing Goethe's reflections on nature in the first issue of *Nature,* and so did Wordsworth in the couplet that served as that journal's motto (p. 69).

Mystical experience is usually regarded as rare, confined to a few saints, sages, and visionaries. But in fact it is surprisingly common. In surveys of random samples of the population in both Britain and the United States, over a third of the people questioned said they had been aware of "a presence or a power" at least once in their lives, and for most of them this experience was very significant.[12] In the thousands of accounts of mystical experience collected by the Religious Experience Research Unit at Oxford, many involve a sense of connection with nature. However, such experience is rarely discussed. Again and again, those who described their experience to the researchers did so with a sense of relief at being able to talk about it. For many, their spiritual or mystical experience seemed to be of supreme importance, but they were unable to discuss it with their families and friends for fear of ridicule or being thought mentally un-balanced. This research has revealed in fact that there is a widespread taboo in our society against admitting to such ex-perience.

> The mood of these accounts is very reminiscent of that which used to surround the public discussion of intimate sexual matters. There is the same feeling of tentativeness, followed by rapid retreat if no response or an insensitive one is detected. . . . Even professional representatives of the sacred are not exempt from the suspicion that they will not understand. There seems to be a feeling that "society" in some way does not give permission for these experi-ences to be integrated into ordinary life.[13]

If nature is inanimate, then the experience of a mystical con-nection with a living presence or power in nature must be illu-sory, and so it is best not to pay too much attention to it lest it have an unbalancing effect on the rational mind. But if nature is alive, such an experience of a living connection may be just what it seems to be.

Remembering Childhood Experience

Many children have in certain moments a mystical sense of their connection with the natural world. Some forget it. Others remember it in a way that serves as a continuing source of inspiration. For example:

> Through the spring, summer and autumn days from about the age of seven, I would sit alone in my little house in the tree tops observing all nature around me and the sky overhead at night. I was too young to be able to think and reason in the true sense, but with the open, receptive mind of a young, healthy boy I slowly became aware of vague, mysterious laws in everything around me. I must have become attuned to nature. I felt these laws of life and movement so deeply they seemed to saturate my whole mind and body, yet they always remained just beyond my grasp and understanding.[14]

This person later became a writer, first on Marxism and later on theosophy. A woman art teacher recalled a more organic, less intellectual response when walking on the Pangbourne Moors at about the age of five:

> Suddenly I seemed to see the mist as a shimmering gossamer tissue and the harebells, appearing here and there, seemed to shine with a brilliant fire. Somehow I understood that this was the living tissue of life itself, in which all that we call consciousness is embedded, appearing here and there as a shining focus of energy in the more diffused whole. In that moment I knew that I had my special place, as had all other things, animate and so-called inanimate, and that we were all part of this universal tissue which was both fragile yet immensely strong, and utterly good and benificent.[15]

Others are primarily impressed by the experience of friendliness. For example, this is how a management consultant remembered a spiritual experience in the early morning near his home when he was five or six:

> The dew on the grass seemed to sparkle like iridescent jewels in the sunlight, and the shadows of the houses and trees seemed friendly and protective. In the heart of the child that I was there suddenly seemed to well up a deep and overwhelming sense of gratitude, a sense of unending peace and security which seemed to be part of the beauty of the morning, the love and protective and living presence which included all that I had ever loved and yet was something much more.[16]

Even if we cannot remember an intuitive sense of connection with nature in childhood, the fact remains that in our formative years we establish patterns of relationship with the natural world that continue to influence us unconsciously. They affect our desire to get back to nature. Sometimes they even shape our subsequent careers.

A few years ago, as I described in the Introduction to this book, I remembered a long-forgotten incident from my own childhood when I saw how a fence made out of willow stakes had grown into a row of vigorous young trees. Much of my scientific career has been concerned with the study of death and regeneration in plants, and this book tells a story not unlike that of the fence that came to life.

Over the past few years, I have been asking some of my professional scientific colleagues about childhood experiences that might have influenced their subsequent interests. Very few had thought about it at all. The two who were most easily able to recall such an experience are both experts on snakes. Both told me that they vividly remembered their first encounter with a snake in the wild as a child; something about it fascinated them and ultimately led them to spend their professional life working with these reptiles.

A colleague in developmental biology, who is much preoccupied with ideas of waves, rhythms, and flows in living organisms, grew up in Canada. He remembered in particular his boyhood passion for canoeing. For him, flows, waves, and rhythms are not only physical processes that can be modeled mathematically; they correspond to living experiences.

One of my colleagues in plant physiology has spent many years studying the geotropic responses of roots, the way they grow downward in response to gravity; he has paid particular attention to the role of starch grains, which by sinking within cells act as sensors for the gravitational field. He isolates such starch grains and other subcellular structures by mashing roots, then filtering the resulting fluid and centrifuging the particles. When we were walking in the Black Forest together, I asked him if he could remember any childhood experiences related to his professional interests. At first all he could think of was a vague youthful desire to press forward the frontiers of science. I then for some reason asked him about his grandmother. He remembered her fondly; as a child, he had lived with her during the war on her farm in Bavaria. He said he particularly remembered the way she grew potatoes and her delicious potato dumplings. He lovingly described how she mashed the potatoes, filtered them through muslin, and then allowed the starch to settle. And this was more or less what he had been doing for over twenty years!

Edward O. Wilson has suggested that it is quite generally the case that scientific innovation is rooted in childhood experience:

> You start by loving a subject. Birds, probability theory, explosives, stars, differential equations, storm fronts, sign language, swallowtail butterflies—the odds are that your obsession will have begun in childhood. The subject will be your lodestar and give sanctuary in the shifting mental universe.

A pioneer in molecular biology once told me that his fascination with the replication of DNA molecules began when he was given an erector set as a child. Playing with the toy, he saw the possibilities of creation by the multiplication and rearrangement of identical units. The great metallurgist Cyril Smith owed his devotion to alloys to the fact that he was color blind. The impairment caused him to turn his attention at an early age to the intricate black-and-white patterns to be seen everywhere in nature, to swirls, filigree, and banding, and eventually to the fine structure of metal. Albert Camus spoke for all such innovators when he said that "a man's work is nothing but this slow trek to rediscover, through the detours of art, those two or three great and simple images in whose presence his heart first opened."[17]

It may not always be possible to relate our interests in later life to experiences in childhood. Nevertheless, I have been impressed again and again by the insights that can flow from bringing to mind such connections, usually long forgotten.

Recognizing Sacred Places

For all of us, particular places are of great personal significance. First and foremost is the place of our birth. Many people feel that their native place is somehow sacred and often want to be buried or have their ashes scattered there. The birthplaces of famous men and women are often visited in a spirit of pilgrimage. And millions of Americans, Australians, and others descended from emigrants make pilgrimages to the old country to reestablish contact with their ancestral homes.

Then there are all the places at which important events have occurred in our lives from childhood on, including those at which we have received moments of illumination and insight or a sense of the numinous, the holy. Such places have a continuing significance for each of us.

Traditional peoples think of their hearths and homes as sacred. They also relate their lives to the recognized sacred places in their locality—temples, shrines, sacred trees, holy wells, churches, cathedrals, mosques, synagogues. For those who experience the places where they live and work as desacralized, devoid of magic and mystery, and want to rediscover this lost dimension, there are several helpful steps. One is to become aware of the local geography and geomancy, the lie of the land, the qualities of the surroundings, and the life of local plants and animals. Another is to learn the stories of the place, to rediscover the local myths and to learn the names of the guardian spirits or patron saints. Yet another is to recognize the local sacred places by visiting them. And most effective of all is to open oneself in prayer to the sacred presence in the place.

Finally, as discussed in Chapter 8, there is the rediscovery of the spirit of pilgrimage—visiting sacred places as pilgrims rather than tourists. As pilgrims, we visit such places in a frame of mind open to their particular power or spirit. And if we do not want to be open to their power, it might be better to stay away.

Many benefits could flow from this transformation. The Japanese, for example, have a traditional sense of the sacredness of their land, maintained in an explicit form through the Shinto religion. They have one of the highest proportions of land under forest of any industrialized country and are averse to destroying this part of their environment. But through their economic demands, the environment of many traditional peoples, such as the forest people of Borneo, is being devastated, and the Japanese fishing and whaling fleets are among the most rapacious in the world. If they recognized the sacredness of the natural world not only within Japan but elsewhere as well, their attitudes would change—and change far more deeply than through political pressure from other countries, most of which also have unimpressive environmental records. And the same goes for all other peoples.

Remembering Sacred Time

Our everyday lives are structured by the cycles of day and night, waking and sleeping. For most people in the world, these cycles are sacralized through daily rituals and prayers; many Hindus, for example, greet the rising sun with the appropriate mantra. Even for those whose sense of time is entirely secular, the patterns of the day and night have their particular qualities. Nights, for example, are more private, more fearful, and more mysterious than days; they are the abode of dreams.

Calendars have a sacred as well as a practical aspect. When the seven-day week was established among the Jewish people, it was essential to its rhythm that the seventh day, the Sabbath, should be a day of rest. The same rhythm is preserved among the Christian and Islamic inheritors of the seven-day cycle, but with Sunday and Friday as the holy days instead of Saturday. And this rhythm remains the basis of our modern life. Hundreds of millions of people still observe this cycle through participating in weekly public worship. Many also remember the spiritual dimension of the weekly cycles of time in a more private way; for example, on Friday evenings, many Jewish people observe the traditional ceremony in which a woman of the house lights the candles and invokes the sabbath bride, or *shekinah*, the feminine presence of God. And for many millions of people, weekends serve as times of spiritual refreshment through "getting back to nature," whether with a conscious religious intention or not.

The linking of calendars to the cycles of the moon and sun reminds us of the celestial context of our earthly life, and seasonal festivals celebrate the quality of the time of the year and give it a sacred dimension. For example, one of the attractions of the Indonesian island of Bali is the enchantment of time and place through the celebration of festivals; tourists are drawn there in their millions. Bali, like other parts of the world not yet disenchanted through modern education and development, evokes a nostalgia in us for something we have lost. But if we are not to live vicariously on other people's cultures and traditions, we need to recover a sense of participation in our own. Seasonal festivals and holy days give us many opportunities to do so.

Gratitude

It is hard to feel a sense of gratitude for an inanimate, mechanical world proceeding inexorably in accordance with eternal laws of nature and blind chance. And this is a great spiritual loss, for it is through gratitude that we acknowledge the living powers on which our own lives depend; through gratitude we enter into a conscious relationship to them; through gratitude we can find ourselves in a state of grace.

All religions provide opportunities for giving thanks, both through simple everyday rituals, like saying grace before meals, and also in collective acts of thanksgiving. These customary expressions of gratitude help to remind us that we have much to be thankful for. Each religion has its own ways of recognizing the living powers on which we all depend and of establishing a relationship to these powers through thanksgiving.

For those for whom traditional religious practices seem empty and meaningless, there are three possibilities: first, to recognize no living power greater than humanity and hence to recognize neither a need for gratitude nor a means of expressing it; second, to feel such gratitude privately but with no means of public expression; third, to find new ways of expressing gratitude collectively and new conceptions of the life-giving powers to whom thanks are due.

The Power of Prayer

Prayer is more than positive thinking, a technique for trying to get what one wants through the power of the mind; it is a form of dialogue with a higher conscious power or powers. Prayers in all religions open with an invocation, naming the power to whom they are addressed. They then generally establish a relationship with this power, recognizing the dependence of the worshiper upon it, and they may then go on to make petitions. Think, for example, of the structure of the Lord's Prayer.

From a humanistic point of view, this is all mere wishful thinking, and if it has any benefits, they are no more than psychological: praying may make people feel better. That alone is no small benefit. However, for those who actually pray—probably the great majority of humanity—the power of prayer is believed to extend far beyond the mind of the person praying. For example, in the 1980s there were a number of international movements that involved millions of people in praying for peace. Those who participated, including myself, believed that such prayers could in some way influence events and collective attitudes in a way that went beyond our individual minds.

Many people who pray find that prayers can be answered in surprising ways. I do myself. But here the skeptic within or without is always ready to invoke the powers of "self-deception" and "random coincidence," to argue that if what was prayed for came to pass, it would have happened anyway. The thawing of the Cold War is a case in point. Some people, including myself, believe that the power of prayer has played some part in it; skeptics believe that it would have happened anyway. Either way, it is a matter of belief or opinion; it is no more possible to prove that prayer played no part than to prove that it did.

We need to respond to our present ecological crisis practically, by making appropriate social, political, economic, and technological changes. We need to look at the attitudes that have led to such devastation of the earth and to find a more harmonious way of living. And those of us who believe in the power of prayer need to pray for forgiveness and guidance. If a wiser and juster human order comes about, if a new harmony develops between humanity and the living world, this would indeed seem like an answer to prayer.

A New Renaissance

As soon as we allow ourselves to think of the world as alive, we recognize that a part of us knew this all along. It is like emerging from winter into spring. We can begin to reconnect our mental life with our own direct intuitive experiences of nature. We can participate in the spirits of sacred places and times. We can see that we have much to learn from traditional societies who have never lost their sense of connection with the living world around them. We can acknowledge the animistic traditions of our ancestors. And we can begin to develop a richer understanding of human nature, shaped by tradition and collective memory, linked to the earth and the heavens, related to all forms of life, and consciously open to the creative power expressed in all evolution. We are reborn into a living world.

NOTES

Introduction

1. Sheldrake (1973).
2. Sheldrake (1974).
3. Sheldrake (1984); Chauhan, Venkataratnam, and Sheldrake (1987).

1   Mother Nature and the Desecration of the World

1. Partridge (1958).
2. Neumann (1963).
3. Eliade (1959), p. 138.
4. Ibid., p. 139.
5. Eliade (1958), pp. 247–49.
6. Levy (1963).
7. Hillman (1979).

8. Quoted by King-Hele (1977), p. 75.
9. Ibid.
10. Merchant (1982).
11. Eliade (1979), pp. 52–55.
12. Gimbutas (1974).
13. Eliade (1958), p. 259.
14. Quoted in Merchant (1982), p. 31.
15. Ibid., p. 8.
16. Gimbutas (1974).
17. Eisler (1987).
18. Turner (1983).
19. Eisler (1987).
20. Graves (1955).
21. Brown et al. (1968), p. 10.
22. Weber (1978), pp. 138–73. For a penetrating essay on the roots of the desanctification of nature in the West, see Sherrard (1987).
23. Hastings (1909), pp. 56, 352–53.
24. Frazer (1918), vol. 3, chap. 15.
25. Beresford Ellis (1985).
26. Quoted in Bentley (1985).
27. Warner (1985).
28. Eire (1986), p. 224.
29. Dickens (1964).
30. Quoted in Aston (1988), p. 6.
31. Quoted in Wall (1905), p. 138.
32. For illuminating discussions, see Roszak (1973) and Berman (1984).
33. Aston (1988).
34. Eire (1986).
35. Eire (1986), p. 207.
36. Walker (1983).
37. Partridge (1958).
38. Shulman (1990).

2   The Conquest of Nature and the Scientific Priesthood

1. E.g., White (1967).
2. Aristotle, *Politics* 1256b (trans. 1941).
3. Ciochon, Olsen, and James (1990).
4. Stuart (1986); Simmons (1988).
5. Quoted in Turner (1983), p. 170.
6. Yates (1964); Thomas (1973).
7. Yates (1979); Berman (1984).
8. Butler (1952).
9. Leiss (1972), p. 51.
10. Ibid., p. 50.
11. Lemmi (1971).
12. E.g., Griffin (1978); Merchant (1982); Keller (1985).
13. Quoted in Merchant (1982), p. 169.
14. Ibid., pp. 168–71.
15. Keller (1985), pp. 53–54.
16. Ibid., p. 54.
17. Collingwood (1945).
18. Gilson (1930), p. 215.
19. Gilson (1984).
20. Gilson (1930).
21. Quoted in Burtt (1932), p. 44.
22. Ibid., p. 48.
23. Lear (1965).
24. Trans. Wallace (1911), p. 80.
25. Descartes (trans. 1985), vol. 1, p. 101.
26. Trans. Wallace (1911), p. 87.
27. For a detailed discussion, see Sheldrake (1988), chap. 5.
28. Thomas (1984).
29. Wallace (1911), p. 81.
30. Descartes (trans. 1985), vol. 1, p. 317.
31. Thomas (1984), p. 34.
32. Ibid., p. 33.

33. Ibid.
34. Driesch (1914).
35. Hazen (1989).
36. Campbell (1956), p. 30.
37. For a stimulating discussion of the development of the idea of scientific objectivity, see Castillejo (1982).
38. Whyte (1979).
39. Descartes (trans. 1985), vol. 1, p. 127.
40. For a far-ranging discussion of the mind-body split, see Berman (1989).
41. Keller (1985).
42. Quoted in Burtt (1932), p. 75.
43. Turner (1983), pp. 266–69.
44. Ibid., pp. 282–83.

3 Returning to Nature

1. Pope, A. (1711), *Essay on Criticism,* lines 68–72.
2. Quoted in Lovejoy (1960), p. 113.
3. Thomas (1984), p. 258.
4. Ibid., p. 257.
5. Ibid., p. 258.
6. Ibid., p. 267.
7. Ibid., p. 267.
8. Ibid., p. 266.
9. Ibid., pp. 268–69.
10. Farmer (1984).
11. Perrin (1986), p. 16.
12. Emerson (1985), pp. 38–39.
13. Thoreau (1988), pp. 314–15.
14. Thoreau (1983), p. 83.
15. Ibid., p. 183.
16. Quoted in Hoagland (1986), pp. 46–47.
17. Perrin (1986), p. 20.
18. Hoagland (1986), p. 48.

19. Thomas (1984), p. 269.
20. W. Wordsworth, *Miscellaneous Sonnets,* part 1, 34.
21. *Nature* (1869), vol. 1, p. 9.
22. Ibid., pp. 10–11.
23. Darwin (1974), p. 23.
24. Mayr (1982).
25. Darwin (1974), p. 23.
26. Quoted and discussed by J. Wilson (1988).
27. E.g., E. O. Wilson (1984).
28. Darwin (1875), pp. 7–8.
29. Darwin (1859), chap. 3.
30. Bergson (1911), p. 110.
31. Monod (1972), p. 119.
32. Ibid., p. 110.
33. Neumann (1963), p. 43.

4 The Reanimation of the Physical World

1. Gilson (1930, 1984).
2. Westfall (1980), p. 505.
3. Ibid., p. 509.
4. Ibid.
5. Quoted in Whittaker (1951), p. 4.
6. Burnet (1930), p. 48.
7. Needham (1962).
8. Zilsel (1957).
9. Ibid., p. 222.
10. Ibid., p. 223.
11. Ibid.
12. Whittaker (1951), chap. 2.
13. Berkson (1974).
14. Nersessian (1984).
15. Ibid., p. 207.
16. Davies (1984), p. 5.
17. Popper and Eccles (1977), pp. 5–7.

18. Harman (1982).
19. Bynum et al. (1981), pp. 122–23.
20. Laplace (1819), p. 4.
21. Popper (1982), pp. 29–31.
22. Prigogine and Stengers (1984).
23. For a good nontechnical introduction to chaos theory, see Gleik (1988).
24. Popper (1982).
25. Gleik (1988).
26. Quoted in Davies (1987).
27. Abraham and Shaw (1984), vol. 1, p. 27.
28. E.g., Waddington (1966); Thom (1975).
29. Carr (1989).
30. For discussions of the reanimation of nature in science, see Cobb and Griffin (1978); Griffin (1988, 1989).

5   The Nature of Life

1. Hildebrand (1988).
2. Driesch (1914).
3. Dawkins (1976).
4. Whitehead (1925), chap. 6.
5. E.g., Sheldrake and Northcote (1968).
6. Alberts et al. (1983), chap. 19.
7. Driesch (1908).
8. Quoted in Lewin (1984).
9. Ibid.
10. Waddington (1966).
11. Thom (1975), p. 320.
12. E.g., Danckwerts (1982).
13. Sheldrake (1988).
14. Tinbergen (1951); Thorpe (1963).
15. Sheldrake (1981), chap. 11; Sheldrake (1988), chap. 9.
16. Lashley (1950).
17. Boycott (1965).

18. Pribram (1971).
19. Sacks (1985).
20. Jung (1959).
21. Marais (1973).
22. Wilson (1971), p. 317.
23. Marais (1973), pp. 119–20.
24. Sheldrake (1988), chap. 13.
25. Ibid., chaps. 14–15.
26. E.g., Koestler (1967); Whyte (1974).
27. E.g., Varela (1979).
28. Capra (1982).

6 Cosmic Evolution and the Habits of Nature

1. For a historical survey, see Mayr (1982).
2. Long (1969).
3. Pagels (1985), p. 11.
4. Hawking (1980).
5. Hawking (1988), p. 60.
6. For a recent defense of the idea of eternal laws of nature, see Barrow (1988).
7. Hawking (1988), p. 9.
8. Pagels (1985).
9. Plotinus (trans. 1964), p. 65.
10. Barrow and Tipler (1986), p. 5.
11. Ibid., p. 16.
12. Ibid., p. 21.
13. Ibid., p. 23.
14. Butler (1878).
15. Sheldrake (1988), p. 15.
16. E.g., Lewis and John (1972).
17. For a more detailed discussion of these and other examples of habit formation in biological evolution, see Sheldrake (1988).
18. Darwin (1875), vol. 2, p. 27.

19. Rensch (1959).
20. Cairns et al. (1988); Hall (1988).
21. Darwin (1875), vol. 2, p. 354.
22. Ibid., p. 359.
23. Ibid., p. 356.
24. Huxley (1959).
25. For summaries of this evidence, see Sheldrake (1985, 1988).
26. Fisher and Hinde (1949).
27. Hardy (1965).
28. Hinde and Fisher (1951).

7   The Earth Comes Back to Life

1. Kelley (1988).
2. Ibid., p. 78.
3. Ibid., p. 109.
4. Lovelock (1988), p. 212.
5. Lovelock (1979), p. ix.
6. Ibid., p. 86.
7. Lovelock (1988), p. 111.
8. Lovelock (1979), p. 11.
9. For a discussion of the mythological aspect of the Gaia hypothesis, see Thompson (1989).
10. In Bunyard and Goldsmith (1988).
11. Lovelock (1988), p. 14.
12. Lindley (1988).
13. E.g., Akasohu (1989).
14. E.g., Skinner and Porter (1987).
15. Stenflo and Vogel (1988).
16. Stothers (1986).

8   Sacred Times and Places

1. Crichton (1987).
2. Quoted in Walker (1983), p. 625.

3. Frazer (1914).
4. Eliade (1958), p. 391.
5. Quoted in Lane (1988), p. 9.
6. For a helpful discussion of the qualities of places, see Pennick (1987).
7. Lethbridge (1980).
8. Partridge (1958), p. 210.
9. Translation in Eliade (1958), p. 367.
10. Ibid., chap. 10.
11. Lane (1988), p. 3.
12. Ashton (1988).
13. See, for example, Eitel (1873); Roosbach (1984); Walters (1988).
14. E.g., Pennick (1987); Devereux et al. (1989).
15. Pennick (1987), p. 142.
16. Ibid., chap. 4.
17. Devereux et al. (1989).

9 The Greening of God

1. See, for example, the discussions by Fox (1983, 1988) and Griffiths (1989).
2. Quoted in Fox (1988), p. 110. See also Hildegard of Bingen (trans. 1985).
3. See, for example, Griffiths (1976, 1982, 1989).
4. Frazer (1914, 1918).
5. Eliade (1964).
6. Halifax (1982), p. 7.
7. Eliade (1979), vol. 1, p. 185.
8. Luna (1986); McKenna (1991).
9. See, for example, Moody (1975) and Ring (1985).
10. Grey (1985), p. 6.
11. I am indebted to Bill Soskin for this idea.
12. Ashton (1988).
13. Ibid.

14. Begg (1985).
15. Quoted in *Devotions to Our Lady of Czestochowa*, Daughters of St. Paul (1981).
16. Ashton (1988).
17. Augustine (trans. 1873).
18. Sherrard (1987), p. 111.
19. Quoted in Fox (1988), p. 126.
20. For an illuminating discussion, see Griffiths (1989).
21. Fox (1988).
22. Clow (1962); McKenzie (1966); Walker (1983).
23. Translated by Fox (1988), p. 123.
24. Ibid., p. 124.
25. Teilhard de Chardin (1965), pp. 288–89.
26. Birch and Cobb (1981), pp. 196–97.

10   Life in a Living World

1. Quoted in Berry (1988), p. 208.
2. Ibid., pp. 208–09.
3. Ibid., p. 209.
4. For a wise essay on this subject, see Ashby (1978).
5. Brundtland et al. (1987), p. 8.
6. Tokar (1988).
7. E.g., Duval and Sessions (1985). The debate between social and deep ecologists can be followed in journals such as *Environmental Ethics* and *The Ecologist*.
8. E.g., Myers (1985).
9. McKibben (1990), p. 211.
10. British Meteorological Office Report, *The Guardian*, November 14, 1989.
11. E.g., Rackham (1986).
12. Hay (1982), chap. 8.
13. Ibid., p. 159.

14. Quoted in Robinson (1983), pp. 31–32.
15. Ibid., p. 32.
16. Ibid., p. 33.
17. Wilson (1984), pp. 65–66.

REFERENCES

Abraham, R. H., and C. D. Shaw. 1984. *Dynamics: The Geometry of Behavior*. Santa Cruz, Calif.: Aerial Press.

Akasofu, S. 1989. The dynamic aurora. *Scientific American* 260 (5), 54–63.

Alberts, B., et al. 1983. *Molecular Biology of the Cell*. New York: Garland.

Anderson, W., and C. Hicks. 1985. *The Rise of the Gothic*. London: Hutchinson.

Aristotle. 1941. *The Basic Works of Aristotle*. Translated by R. McKeon. New York: Random House.

Ashby, E. 1978. *Reconciling Man with the Environment*. Stanford, Calif.: Stanford University Press.

Ashton, J. 1988. *Mother of Nations: Visions of Mary*. Basingstoke, England: The Lamp Press.

Aston, M. 1988. *England's Iconoclasts*. Vol. 1: *Laws Against Images*. Oxford: Oxford University Press.

Augustine, A. 1873. *On the Trinity: The Works of Aurelius Augustine*. Vol. 7. Translated. Edinburgh: T. and T. Clark.

Bacon, F. 1951. *The Advancement of Learning and New Atlantis*. London: Oxford University Press.

Barnett, S. A. 1981. *Modern Ethology*. Oxford: Oxford University Press.

Barrow, J. D. 1988. *The World within the World*. Oxford: Clarendon Press.

Barrow, J. D., and F. J. Tipler. 1986. *The Anthropic Cosmological Principle*. Oxford: Oxford University Press.

Begg, E. 1985. *The Cult of the Black Virgin*. London: Arkana.

Bentley, J. 1985. *Restless Bones: The Story of Relics*. London: Constable.

Beresford, Ellis P. 1985. *Celtic Inheritance*. London: Muller.

Bergson, H. 1911. *Creative Evolution*. London: Macmillan.

Berkson, W. 1974. *Fields of Force*. London: Routledge and Kegan Paul.

Berman, M. 1984. *The Reenchantment of the World*. New York: Bantam.

———. 1989. *Coming to Our Senses: Body and Spirit in the Hidden History of the West*. New York: Simon & Schuster.

Berry, T. 1988. *The Dream of the Earth*. San Francisco: Sierra Club Books.

Birch, C., and J. B. Cobb. 1981. *The Liberation of Life: From the Cell to the Community*. Cambridge: Cambridge University Press.

Bloxham, J., and D. Gubbins. 1985. The secular variation of the Earth's magnetic field. *Nature* 317, 777–81.

Bord, J., and C. Bord. 1985. *Sacred Waters: Holy Wells and Water Lore in Britain and Ireland*. London: Granada.

Boycott, B. B. 1965. Learning in the octopus. *Scientific American* 212(3), 42–50.

Brown, J. A., J. A. Fitzmyer, and R. E. Murphy, eds. 1968. *The Jerome Biblical Commentary*. Englewood Cliffs, N.J.: Prentice-Hall.

Brundtland, G. H. et al. 1987. *Our Common Future: The World Commission on Environment and Development*. Oxford: Oxford University Press.

Bunyard, P., and E. Goldsmith, eds. 1988. *Gaia, the Thesis, the Mechanisms and the Implications*. Camelford, England: Wadebridge Ecological Centre.

Burnet, J. 1930. *Early Greek Philosophy*. London: Black.

Burtt, E. A. 1932. *The Metaphysical Foundations of Modern Science*. London: Kegan Paul, Trench and Trubner.

Butler, E. M. 1952. *The Fortunes of Faust*. Cambridge: Cambridge University Press.

Butler, S. 1878. *Life and Habit*. London: Cape.

Bynum, W. F., E. J. Browne, and R. Porter, eds. 1981. *Dictionary of the History of Science*. London: Macmillan.

Cairns, J., J. Overbaugh, and S. Miller. 1988. The origin of mutants. *Nature* 335, 142–45.

Campbell, J. 1956. *Hero with a Thousand Faces*. Cleveland, Ohio: Meridian Books.

Capra, F. 1982. *The Turning Point*. London: Wildwood House.

Carr, B. 1989. The dark matter problem. In *Cosmic Perspectives,* ed. V. Vishveshwara. Cambridge: Cambridge University Press.

Castillejo, D. 1982. *The Formation of Modern Objectivity*. Madrid: Ediciones de Arte y Bibliofilia.

Chauhan, Y. S., N. Venkataratnam, and A. R. Sheldrake. 1987. Factors affecting growth and yield of short-duration pigeonpea and its potential for multiple harvests. *Journal of Agricultural Science* 109, 519–29. Cambridge.

Ciochon, R., J. Olsen, and J. James. 1990. *Other Origins: The Search for the Giant Ape in Human Prehistory*. New York: Bantam.

Clow, W. M., ed. 1962. *The Bible Reader's Encyclopaedia and Concordance*. London: Collins.

Cobb, J. B., and D. R. Griffin. 1978. *Mind in Nature: Essays on the Interface of Science and Philosophy*. Washington, D.C.: University Press of America.

Collingwood, R. G. 1945. *The Idea of Nature*. Oxford: Oxford University Press.

Crichton, R. 1987. *Who is Santa Claus?* London: Canongate.

Danckwerts, P. V. 1982. Letter. *New Scientist* 96, 380–81.

Darwin, C. 1859. *On the Origin of Species*. London: Murray.

———. 1875. *The Variation of Animals and Plants Under Domestication*. London: Murray.

———. 1974. *Autobiography*. Oxford: Oxford University Press.

Davies, P. C. W. 1984. *Superforce*. London: Heinemann.

———. 1987. *The Cosmic Blueprint*. London: Heinemann.

Dawkins, R. 1976. *The Selfish Gene*. Oxford: Oxford University Press.

Descartes, R. 1985. *The Philosophical Writings of Descartes*. Translated by J. Cottingham, R. Stoothof, and D. Murdoch. Cambridge: Cambridge University Press.

Devereux, P., J. Steele, and D. Kubrin. 1989. *Earthmind.* New York: Harper and Row.

Dickens, A. G. 1964. *The English Reformation.* London: Batsford.

Driesch, H. 1908. *Science and Philosophy of the Organism.* London: Black.

————. 1914. *The History and Theory of Vitalism.* London: Macmillan.

Duval, B., and G. Sessions. 1985. *Deep Ecology: Living As If Nature Mattered.* Salt Lake City, Utah: Gibbs Smith.

Eire, C.M.N. 1986. *War Against the Idols: The Reformation of Worship from Erasmus to Calvin.* Cambridge: Cambridge University Press.

Eisler, R. 1987. *The Chalice and the Blade.* San Francisco: Harper and Row.

Eitel, J. 1873. *Feng Shui.* London: Trubner.

Eliade, M. 1958. *Patterns in Comparative Religion.* London: Sheed and Ward.

————. 1959. *The Sacred and The Profane: The Nature of Religion.* New York: Harcourt, Brace and World.

————. 1964. *Shamanism: Archaic Techniques of Ecstasy.* Princeton, N. J.: Bollingen.

————. 1979. *A History of Religious Ideas.* London: Collins.

Emerson, R. W. 1985. *Selected Essays.* Harmondsworth, England: Penguin Books.

Evans, H. E. 1987. Remembering pioneer naturalists. In *On Nature,* ed. D. Halpern. San Francisco: North Point Press.

Farmer, A. 1984. *Hampstead Heath.* Barnet, England: Historical Publications.

Finucane, R. C. 1977. *Miracles and Pilgrims: Popular Beliefs in Medieval England.* London: Dent.

Fisher, J., and R. A. Hinde. 1949. The opening of milk bottles by birds. *British Birds,* 42, 347–57.

Fludd, R. 1619. *Utriusque Cosmi Maioris Scilicet. Tomus Secundus De Supernaturali, Naturali, Praeternaturali et Contranaturali. Microcosmi Historia.* Oppenheim: J. T. de Bry.

Fox, M. 1983. *Original Blessing.* Santa Fe, N. M.: Bear and Company.

————. 1988. *The Coming of the Cosmic Christ.* New York: Harper and Row.

Frazer, J. 1914. *The Golden Bough*. London: Macmillan.
————. 1918. *Folk-Lore in the Old Testament*. London: Macmillan.

Gilson, E. 1930. *The Philosophy of St. Thomas Aquinas*. New York: Dorset Press.
————. 1984. *From Aristotle to Darwin and Back Again*. South Bend, Ind.: University of Notre Dame Press.
Gimbutas, M. 1974. *Gods and Goddesses of Old Europe*. London: Thames and Hudson.
Gleick, J. 1988. *Chaos: Making a New Science*. London: Heinemann.
Graves, R. 1955. *The Greek Myths*. Harmondsworth, England: Penguin.
Grey, M. 1985. *Return from Death. An Exploration of the Near-Death Experience*. London: Arkana.
Griffin, D. R., ed. 1988. *The Reenchantment of Science: Postmodern Proposals*. Albany, N.Y.: State University of New York Press.
Griffin, D. R. 1989. *God and Religion in the Postmodern World*. Albany, N.Y.: State University of New York Press.
Griffin, S. 1978. *Woman and Nature*. New York: Harper and Row.
Griffiths, B. 1976. *Return to the Centre*. London: Collins.
————. 1982. *The Marriage of East and West*. London: Collins.
————. 1989. *A New Vision of Reality*. London: Collins.

Haeckel, E. 1892. *The History of Creation*. London: Kegan Paul.
————. 1910. *The Evolution of Man*. London: Watts.
Halifax, J. 1982. *Shaman: The Wounded Healer*. London: Thames and Hudson.
Hall, B. G. 1988. Adaptive evolution that requires multiple spontaneous mutations. *Genetics* 120, 887–97.
Hardy, A. 1965. *The Living Stream*. London: Collins.
Harman, P. M. 1982. *Energy, Force and Matter: The Conceptual Development of Nineteenth-Century Physics*. Cambridge: Cambridge University Press.
Hastings, J. S., ed. 1909. *Dictionary of the Bible*. Edinburgh: Clark.
Hawking, S. 1980. *Is the End in Sight for Theoretical Physics?* Cambridge: Cambridge University Press.
————. 1988. *A Brief History of Time*. London: Bantam.
Hay, D. 1982. *Exploring Inner Space*. Harmondsworth, England: Penguin.
Hazen, R. 1989. Battle of the supermen. *The Guardian* (15 April). London.

Hildebrand, M. von. 1988. An Amazonian tribe's view of cosmology. In *Gaia, the Thesis, the Mechanisms and the Implications,* eds. P. Bunyard and E. Goldsmith. Camelford, Cornwall: Wadebridge Ecological Centre.

Hildegard of Bingen. 1985. *Illuminations of Hildegard of Bingen.* Translated. Santa Fe, N. M.: Bear and Company.

Hillman, J. 1979. *The Dream and the Underworld.* New York: Harper and Row.

Hinde, R. A., and J. Fisher. 1951. Further observations on the opening of milk bottles by birds. *British Birds* 44, 393–96.

Hoagland, E. 1986. In praise of John Muir. In *On Nature,* ed. D. Halpern. San Francisco: North Point Press.

Hope, R. C. 1893. *The Legendary Lore of the Holy Wells of England.* London: Elliot Stock.

Huxley, F. 1959. Charles Darwin: Life and habit. *The American Scholar* (Fall/Winter), 1–19.

Jung, C. G. 1959. *The Archetypes and the Collective Unconscious.* London: Routledge and Kegan Paul.

Kahn, F. 1949. *The Secret of Life: The Human Machine and How it Works.* London: Odhams.

Keller, E. F. 1985. *Reflections on Gender and Science.* New Haven, Conn.: Yale University Press.

Kelley, K. W., ed. 1988. *The Home Planet.* Reading, Mass.:Addison-Wesley.

King-Hele, D. 1977. *Doctor of Revolution: The Life and Genius of Erasmus Darwin.* London: Faber and Faber.

Kirk, G. S., and J. E. Raven. 1957. *The Presocratic Philosophers.* Cambridge: Cambridge University Press.

Koestler, A. 1967. *The Ghost in the Machine.* London: Hutchinson.

Lane, B. C. 1988. *Landscapes of the Sacred: Geography and Narrative in American Spirituality.* New York: Paulist Press.

Laplace, P. S. 1819. *A Philosophical Essay on Probabilities.* Reprint, New York: Dover, 1951.

Lashley, K. l950. In search of the engram. *Symposia of the Society for Experimental Biology* 4, 454–83.

Lear, J. 1965. *Kepler's Dream.* Berkeley, Calif.: University of California Press.

Leiss, W. 1972. *The Domination of Nature.* Boston: Beacon Press.

Lemmi, C. W. 1971. *The Classic Deities in Bacon*. New York: Octagon Books.

Lethbridge, T.D.C. 1980. *The Essential T. C. Lethbridge*. London: Routledge and Kegan Paul.

Levy, G. R. 1963. *The Gate of Horn*. London: Faber and Faber.

Lewin, R. 1984. Why is development so illogical? *Science* 224, 1327.

Lewis, K. R., and B. John. 1972. *The Matter of Mendelian Heredity*. London: Longman.

Lindley, D. 1988. Is the Earth alive or dead? *Nature* 332, 483–84.

Long, C. H. 1969. *Alpha: The Myths of Creation*. New York: Collier Books.

Lovejoy, A. O. 1960. *Essays in the History of Ideas*. New York: Capricorn Books.

Lovelock, J. 1979. *Gaia: A New Look at Life on Earth*. Oxford: Oxford University Press.

———. 1988. *The Ages of Gaia: A Biography of Our Living Earth*. Oxford: Oxford University Press.

Luna, L. E. 1986. *Vegetalismo: Shamanism among the Mestizo Population of the Peruvian Amazon*. Stockholm, Sweden: Almqvist and Wicksell.

Marais, E. 1973. *The Soul of the White Ant*. Harmondsworth, England: Penguin.

Mayr, E. 1982. *The Growth of Biological Thought*. Cambridge, Mass.: Harvard University Press.

McKenna, T. 1991. *Plants, Drugs and History*. New York: Bantam.

McKenzie, J. L. 1966. *Dictionary of the Bible*. London: Chapman.

McKibben, B. 1990. *The End of Nature*. London: Viking.

Merchant, C. 1982. *The Death of Nature: Women, Ecology and the Scientific Revolution*. London: Wildwood House.

Metzner, R., ed. 1989. *Gaia Consciousness: The Re-Emergent Goddess and the Living Earth*. San Francisco: Green Earth Foundation.

Monod, J. 1972. *Chance and Necessity*. London: Collins.

Moody, R. 1975. *Life After Life*. Atlanta, Ga.: Mockingbird Books.

Morgan, T. H. 1901. *Regeneration*. New York: Macmillan.

Myers, N., ed. 1985. *The Gaia Atlas of Planet Management*. London: Pan Books.

Needham, J. 1962. *Science and Civilization in China*. Vol. 4, part 1. Cambridge: Cambridge University Press.

Nersessian, N. J. 1984. Aether/Or: the creation of scientific concepts. *Studies in the History and Philosophy of Science* 15, 175–212.

Neumann, E. 1963. *The Great Mother: An Analysis of the Archetype.* Princeton, N.J.: Princeton University Press.

Nodier, J. E. C., and J. Taylor. 1845. *Voyages Pittoresques et Romantiques dans l'ancienne France.* Paris.

Pagels, H. R. 1985. *Perfect Symmetry.* London: Joseph.

Partridge, E. 1958. *Origins: A Short Etymological Dictionary of Modern English.* London: Routledge and Kegan Paul.

Pennick, N. 1979. *The Ancient Science of Geomancy.* London: Thames and Hudson.

———. 1987. *Earth Harmony.* London: Century.

Perrin, N. 1986. Forever Virgin: The American View of America. In *On Nature,* ed. D. Halpern. San Francisco: North Point Press.

Plato. 1955. *The Republic.* Translation. Harmondsworth, England: Penguin.

Plotinus. 1964. *The Essential Plotinus.* Translation. New York: Mentor.

Popper, K. R. 1982. *The Open Universe: An Argument for Indeterminism.* London: Hutchinson.

Popper, K. R., and J. C. Eccles 1977. *The Self and Its Brain.* Berlin, Germany: Springer.

Pribram, K. 1971 *Languages of the Brain.* Englewood Cliffs, N.J.: Prentice-Hall.

Prigogine, I., and I. Stengers. 1984. Order Out Of Chaos. London: Heinemann.

Rackham, O. 1986. *The History of the Countryside.* London: Dent.

Rensch, B. 1959. *Evolution Above the Species Level.* London: Methuen.

Ring, K. 1985. *Heading Toward Omega: In Search of the Meaning of the Near-Death Experience.* New York: Morrow.

Robinson, E. 1983. *The Original Vision: A Study of the Religious Experience of Childhood.* New York: Seabury Press.

Roosbach, S. 1984. *Feng Shui.* London: Hutchinson.

Roszak, T. 1973. *Where the Wasteland Ends: Politics and Transcendence in Post Industrial Society.* London: Faber and Faber.

Sacks, O. 1985. *The Man Who Mistook His Wife for a Hat.* London: Duckworth.

Sheldrake, R. 1973. The production of auxin in higher plants. *Biological Reviews* 48, 509–59.

———. 1974. The ageing, growth and death of cells. *Nature* 250, 381–85.

———. 1981. *A New Science of Life: The Hypothesis of Formative Causation.* New edition, London: Blond and Briggs, 1985.

———. 1984. Pigeonpea physiology. In *The Physiology of Tropical Crops,* ed. P. R. Goldsworthy. Oxford: Blackwell.

———. 1988. *The Presence of the Past: Morphic Resonance and the Habits of Nature.* London: Collins.

Sheldrake, R., and D. H. Northcote. 1968. The production of auxin by tobacco internode tissues. *New Phytologist* 67, 1–13.

Sherrard, P. 1987. *The Rape of Man and Nature.* Ipswich, England: Golgonooza Press.

Shulman, S. 1990. Global change. *Nature* 343, 398.

Simmons, A. H. 1988. Extinct pygmy hippopotamus and early man in Cyprus. *Nature* 333, 554–57.

Skinner, B. J., and S. C. Porter. 1987. *Physical Geology.* New York: Wiley.

Stenflo, J. O., and M. Vogel. 1988. Global resonances in the evolution of solar magnetic fields. *Nature* 319, 285–90.

Stothers, R. B. 1986. Periodicity of the Earth's magnetic reversals. *Nature* 322, 444–46.

Stuart, A. 1986. Who (or what) killed the giant armadillo? *New Scientist* July 17, 29–32.

Teilhard de Chardin, P. 1965. *The Phenomenon of Man.* London: Collins.

Thom, R. 1975. *Structural Stability and Morphogenesis.* Reading, Mass.: Benjamin.

Thomas, K. 1973. *Religion and the Decline of Magic.* London: Penguin.

———. 1984. *Man and the Natural World: Changing Attitudes in England 1500–1800.* Harmondsworth, England: Penguin Books.

Thompson, W. I. 1989. *Imaginary Landscapes: Making Worlds of Myth and Science.* New York: St. Martin's Press.

Thoreau H D. 1983. *Walden.* London: Penguin Books.

———. 1988. *The Maine Woods.* London: Penguin Books.

Thorpe, W. H. 1963. *Learning and Instinct in Animals.* London: Methuen.

Tinbergen, N. 1951. *The Study of Instinct.* Oxford: Oxford University Press.

Tokar, B. 1988. Social ecology, deep ecology and the future of green political thought. *The Ecologist* 18, 132–41.

Turner, F. 1983. *Beyond Geography: The Western Spirit Against the Wilderness.* New Brunswick, N.J.: Rutgers University Press.

Varela, F. J. 1979. *Principles of Biological Autonomy.* New York: North Holland.

Waddington, C. H. 1966. Fields and Gradients. In *Major Problems in Developmental Biology,* ed. M. Locke. New York: Academic Press.

Walker; B. G. 1983. *The Woman's Encyclopedia of Myths and Secrets.* San Francisco: Harper and Row.

Wall, J. C. 1905. *Shrines of British Saints.* London: Methuen.

Wallace, W. 1911. Descartes. *Encyclopaedia Britannica* 11th ed. Vol. 8. Cambridge: Cambridge University Press.

Walters, D. 1988. *Feng Shui.* London: Pagoda.

Warner, M. 1976. *Alone of All Her Sex: The Myth and Cult of the Virgin Mary.* London: Weidenfeld and Nicolson.

———. 1985. *Monuments and Maidens: The Allegory or the Female Form.* London: Pan.

Weber, M. 1978. *Selections in Translation,* ed. W. G. Runciman. Cambridge: Cambridge University Press.

Westfall, R. S. 1980. *Never at Rest: A Biography of Isaac Newton.* Cambridge: Cambridge University Press.

White, L. 1967. The historical roots of our ecologic crisis. *Science* 155, 1203–07.

Whitehead, A. N. 1925. *Science and the Modern World.* New York: Macmillan.

Whittaker, E. 1951. *A History of the Theories of Aether and Electricity.* London: Nelson.

Whyte, L. L. 1914. *The Universe of Experience.* New York: Harper and Row.

———. 1979. *The Unconscious Before Freud.* London: Friedmann.

Wilson, E. O. 1971. *The Social Insects.* Cambridge, Mass.: Harvard University Press.

———. 1984. *Biophilia.* Cambridge, Mass.: Harvard University Press.

Wilson, J. 1988. Nausea, or how Darwin became a machine. *Harvest* 34, 132–41.

Wolpert, L., and J. Lewis. 1975. Towards a theory of development. *Federation Proceedings* 34, 14–20.

Yates, F. A. 1964. *Giordano Bruno and the Hermetic Tradition.* London: Routledge and Kegan Paul.
———. 1979. *The Occult Philosophy in the Elizabethan Age.* London: Routledge and Kegan Paul.

Zilsel, E. 1957. The origins of Gilbert's scientific method. In *Roots of Scientific Thought,* eds. P. P. Wiener and A. Noland. New York: Basic Books.

INDEX

Figures and illustrations have page references in italics.

Adam, 41
Addison, Joseph, 62
Aeschylus, 13
Agriculture, 15, 17–18
  festivals, 24
  and Jews, 24
  see also Pastoralists
Alaska, 58
Aleksandrov, Aleksandr, 150
All Saints' Day, 167
All Souls' Day, 167
Amazon
  forest devastation, 41, 58, 156
  forest people, 187
  psychedelic plants, 187
Amber, 83
America
  conquest of, 38
  desacralization of, 59, 65
  development vs. conservation in, 66
  discovery of, 171
  mapping of, 59–60
  national parks, 68
  Thanksgiving, 169
  Western expansion, 58–59
  wilderness in, 65–68
Animals
  Biblical creation of, 20
    naming of, 41
  birds
    blind (swiftlet), 139–140
    Cosmic, 186
    domestic fowl, 142
    dunlin, 119
    migration, 213
    ostriches, 142
    tit, 144
  buffalo, 58
  Descartes and mechanistic view of, 50–53
  domesticated, 142
  extinctions, 36, 41, 155, 162
    dinosaurs, 155
    giant armadillo, 36
    mammoths, 36
    pygmy hippopotamus, 36

feral
  cats, 139
  pigs, 139
fish, shoals of, 119
flying phalanger, *140*
flying squirrel, *140*
fowl, 142
habit and, 142
human relationship with, 211–212
instinct, 113–115
jerboas, *140*
marsupials, 138–139, *140*
moles, *140*
mud wasp, 113, *114*, 115
placental mammals, 138–139, *140*
rat, 115
reindeer, 119
sacrifice, 179–180
social organization of, 117–119
soul in, 50–53
swiftlet, 139, 141
tits, 143–144
vitalists, 53
vivisection, 53
whale, 119
wild type, 136, 138
Anima mundi, 80–81, 82, 85
Animism, 4–5, 92–93, 151–153
  ancient Greece and, 44, 88
  fairy tales and, 4
  and the feminine, 4, 199
  in Judaism and Christianity, 184–185
  matter and, 88
  medieval Europe, 44–45, 183
  new, 157, 206–207
  vitalism and, 99–101, 103–105, *106*
  see also Nature; Soul; Vitalism
Anthropic principle, 158
  final, 133
  strong form, 132
  weak form, 132
Apocalypse, 207–208
Apollo, 19, 37
Aristotle, 35, 80, 202
  entelechy, 99, 105